Reinventing Manufacturing and Business Processes Through Artificial Intelligence

Innovations in Big Data and Machine Learning

Series Editors: Rashmi Agrawal and Neha Gupta

This series will include reference books and handbooks that will provide the conceptual and advanced reference materials that cover building and promoting the field of Big Data and Machine Learning which will include theoretical foundations, algorithms and models, evaluation and experiments, applications and systems, case studies, and applied analytics in specific domains or on specific issues.

Artificial Intelligence and Internet of Things
Applications in Smart Healthcare
Edited by Lalit Mohan Goyal, Tanzila Saba, Amjad Rehman, and Souad Larabi

Reinventing Manufacturing and Business Processes Through Artificial Intelligence
Edited by Geeta Rana, Alex Khang, Ravindra Sharma, Alok Kumar Goel, and Ashok Kumar Dubey

For more information on this series, please visit: www.routledge.com/Innovations-in-Big-Data-and-Machine-Learning/book-series/CRCIBDML

Reinventing Manufacturing and Business Processes Through Artificial Intelligence

Edited by Geeta Rana, Alex Khang,
Ravindra Sharma, Alok Kumar Goel,
and Ashok Kumar Dubey

CRC Press
Taylor & Francis Group
Boca Raton London New York

CRC Press is an imprint of the
Taylor & Francis Group, an **informa** business

First edition published 2022
by CRC Press
6000 Broken Sound Parkway NW, Suite 300, Boca Raton, FL 33487–2742

and by CRC Press
4 Park Square, Milton Park, Abingdon, Oxon, OX14 4RN

© 2022 selection and editorial matter, Geeta Rana, Alex Khang, Ravindra Sharma, Alok Kumar Goel, and Ashok Kumar Dubey; individual chapters, the contributors

CRC Press is an imprint of Taylor & Francis Group, LLC

Reasonable efforts have been made to publish reliable data and information, but the author and publisher cannot assume responsibility for the validity of all materials or the consequences of their use. The authors and publishers have attempted to trace the copyright holders of all material reproduced in this publication and apologize to copyright holders if permission to publish in this form has not been obtained. If any copyright material has not been acknowledged please write and let us know so we may rectify in any future reprint.

Except as permitted under U.S. copyright law, no part of this book may be reprinted, reproduced, transmitted, or utilized in any form by any electronic, mechanical, or other means, now known or hereafter invented, including photocopying, microfilming, and recording, or in any information storage or retrieval system, without written permission from the publishers.

For permission to photocopy or use material electronically from this work, access www. copyright.com or contact the Copyright Clearance Center, Inc. (CCC), 222 Rosewood Drive, Danvers, MA 01923, 978–750–8400. For works that are not available on CCC please contact mpkbookspermissions@tandf.co.uk

Trademark notice: Product or corporate names may be trademarks or registered trademarks and are used only for identification and explanation without intent to infringe.

Library of Congress Cataloging-in-Publication Data
Names: Rana, Geeta, editor.
Title: Reinventing manufacturing and business processes through artificial intelligence / edited by Geeta Rana, Alex Khang, Ravindra Sharma, Alok Kumar Goel, and Ashok Kumar Dubey.
Description: First edition. | Boca Raton : CRC Press, 2022. | Series: Innovations in big data and machine learning | Includes bibliographical references and index.
Identifiers: LCCN 2021031081 (print) | LCCN 2021031082 (ebook) | ISBN 9780367702090 (hbk) | ISBN 9780367702106 (pbk) | ISBN 9781003145011 (ebk)
Subjects: LCSH: Industry 4.0. | Production engineering. | Automation. | Artificial intelligence.
Classification: LCC T59.6 .R45 2022 (print) | LCC T59.6 (ebook) | DDC 658.4/038028563—dc23
LC record available at https://lccn.loc.gov/2021031081
LC ebook record available at https://lccn.loc.gov/2021031082

ISBN: 978-0-367-70209-0 (hbk)
ISBN: 978-0-367-70210-6 (pbk)
ISBN: 978-1-003-14501-1 (ebk)

DOI: 10.1201/9781003145011

Typeset in Times
by Apex CoVantage, LLC

Contents

Preface..vii

Acknowledgments..ix

Editors...xi

Contributors xiii

Chapter 1   The Role of Artificial Intelligence in Adopting Green
HRM Practices 1

Minisha Gupta

Chapter 2   The Role of Artificial Intelligence in Blockchain Applications.........19

Haru Hong Khanh and Alex Khang PH

Chapter 3   The Rise of Artificial Intelligence in Modern Healthcare Sector......39

Yogesh Pant and Balaji Dhanasekaran

Chapter 4   Artificial Intelligence in Manufacturing 63

V. Harish, D. Krishnaveni, and A. Mansurali

Chapter 5   Customer Behavior Prediction for E-Commerce Sites Using
Machine Learning Techniques: An Investigation 79

*G. Rajesh, S.P. Preethi, R. Shanmuga Priya, L. Rajesh, and
X. Mercilin Raajini*

Chapter 6   The Impact of Artificial Intelligence on Global Business
Practices 95

Bhakti Parashar and Geeta Rana

Chapter 7   Road Map for Implementation of IoT in Metal Cutting for
the Process Monitoring System to Improve Productivity 115

Mukhtar Sama and Ashwini Kumar Saini

Chapter 8   Reinventing HR with Conversational Artificial Intelligence:
A Conceptual Framework 137

Neetu Kumari and Geeta Rana

v

Chapter 9 AI and Business Sustainability: Reinventing Business
Processes 153

D. Krishnaveni, V. Harish, and A. Mansurali

Index 173

Preface

The scope of artificial intelligence (AI) in business transformation is constantly growing, and there are no signs of it coming to a halt anytime soon. At some point in time soon, any organization failing to capitalize on AI might not be able to stay relevant or competitive in the market. In business, artificial intelligence has a wide range of uses. In fact, most of us interact with artificial intelligence in some form or another on a daily basis. From the mundane to the breathtaking, artificial intelligence is already disrupting virtually every business process in every industry. The future is definitely gravitating toward automation. Artificial intelligence will be the driving force behind eliminating the human error factor from business operations. Extensive and complex datasets are already being analyzed within a matter of minutes, and useful insights can be churned out more easily. AI has already changed the way we do business, and it is going to accelerate operations in more innovative ways that will benefit entrepreneurs in the long run. As artificial intelligence technologies proliferate, they are becoming an imperative for businesses that want to maintain a competitive edge. This technology provides significant development opportunities that many businesses and companies have already been fast to grab. Many businesses are disrupting their industries with applications of AI that not only bring on change but also bring in different types of organizational practices.

In this book, we will look at how artificial intelligence is beneficial to businesses. This book offers a practical guide to AI techniques that are used in business. The book does not focus on AI models and algorithms, but instead provides an overview of the most popular and frequently used models and techniques in business. This allows the book to easily explain AI paradigms and concepts for business students and executives.

The goal of this book is to focus on concrete tips and methods to help prototype, iterate and deploy models. This book is an anthology of 9 chapters that provide a rich repertoire of tools and techniques across business functions researched, tested and validated in various business settings. This book and its chapters structure is focused on helping you drive concrete business decisions through applications of artificial intelligence. This book will also enable readers to learn how to use multi-functional area tools, techniques, innovative frameworks, practices and approaches for understanding, assessing and managing the strategic value drivers of business excellence. Overall, this book brings forth a new stream of thoughts by a few fine researchers in the domain of artificial intelligence. We wish and hope that this book will generate interest among not only fellow academicians, but also engineering and management practitioners.

Happy reading!

Geeta Rana
Alex Khang
Ravindra Sharma
Alok Kumar Goel
Ashok Kumar Dubey

Acknowledgments

Artificial intelligence (AI) is being widely recognized as the power that will fuel the future global digital economy. AI in the past few years has gained geostrategic importance, and a large number of countries are striving hard to stay ahead with their policy initiatives to get their countries ready. We express our sincere gratitude to our students, colleagues and the executive participants of numerous management development programs who have helped us clarify our concepts of AI; distinguished scholars and authors, whose works we have used over the years in our teaching, research and training, and have thereby become an unconscious part of our ideas and thoughts discussed in this book. The book in its present form has been possible due to keen interest shown by academic colleagues from around the world. We would like to express our gratitude and appreciate the contributions of the authors.

A large number of our friends and colleagues from academia and industry offered their support to make this project successful. This support has come in many forms: review and re-review of chapters, suggestions for revisions and modifications and critical comments to make the chapters more thorough and meaningful. We acknowledge their tremendous support and are thankful for their valuable comments.

We thankfully acknowledge all the support, inspiration and motivation we received from our faculty colleagues. We would remain indebted to their outstanding intellectual efforts.

Our special thanks to Prof. Renu Rastogi and Prof. Santosh Rangnekar, Indian Institute of Technology, Roorkee, India, for their constant encouragement and support in this endeavor. The book in its present shape has been possible due to keen interest shown by academic colleagues from around the world. We acknowledge their tremendous support and are thankful for their valuable comments. We thank our prospective readers in advance, for they would be a source of improvement and further development of this book.

We express our thankful to our publisher, CRC Press (Taylor & Francis Group), and the entire editorial team who lent their wonderful support throughout and ensured timely processing of the manuscript and bringing out of the book.

Finally, we would fail in our duty if we did not acknowledge the most loving care and cooperation enjoyed by us, all through the working on this book, from our dear families for unstained support provided.

Editors

Ashok Kumar Dubey has been professor and dean in the Dept. of Management and Commerce for 18 years at Bangalore University, India. Prof. Dubey is fellow member of three professional institutes, completed Ph.D. from Delhi University and postgraduate work in both management and commerce. He also served 26 years in industry. His research is situated in the field of finance and accounting, with a special focus on forensic and behavioral, academic entrepreneurship and collaborative research and development in industries and academia. He is author of more than 150 research papers in national and international referred journals and more.

Alok Kumar Goel is working at CSIR-Human Resource Development Centre, Ghaziabad, India. He pursued a Ph.D. in the area of knowledge management and human capital creation from Indian Institute of Technology Roorkee. Dr. Goel is recipient of the highly commended award of the 2013 Emerald/EFMD Outstanding Doctoral Research Awards in the knowledge management category. He has published 14 papers in international journals and 27 papers in the proceedings of leading international conferences. Dr. Goel is a visiting scholar of Hasselt University, Belgium wherein he pursued post-doctoral research work in the interdisciplinary areas of knowledge management, open innovation and entrepreneurship. His research interests include knowledge management, digital transformation and human resource management.

Alex Khang has 15 years of teaching information technology and database technology in universities of technology in Vietnam, and over 25 years of non-stop working in the field of software productions and specialized in data engineering for foreign corporations from Sweden, the United States, Singapore, and multinationals. Prof. Khang is a specialist in data engineering and artificial intelligence at IT Corporation and also in the contribution stage of knowledge and experience into the scope of tech talk, a consultant, a part-time lecturer and an evaluator on the sides of content and engineering for the research and editing of master's/Ph.D. theses for local and international institutes and schools of technology.

Ravindra Sharma is assistant professor in the Himalayan School of Management Studies at Swami Rama Himalayan University, Dehradun, India. He has more than 15 years of corporate and academic experience. He holds master's degrees in business administration (MBA) and computer applications (MCA). He also qualified for the University Grant Commission National Eligibility Test in Management (UGC-NET). He has conducted many FDPs for international and national audiences. He has published research papers in referred journals of Emerald, Sage, Springer, IGI Global and Inderscience. He has published books in the area of Internet of Things (IoT) and employer branding with reputed publishers like Taylor & Francis Group, Nova Science Publishers and IRP Publication House. He also contributed chapters in different books published by Springer IGI Global, Taylor & Francis Group, and Palgrave

Macmillan. He has attended various workshops, FDPs and MDPs. He has presented around 28 research papers in national and international conferences. He has also attended a paper development workshop in the Indian Institute of Management Rohtak, India. He has been honored as a session chair and keynote speaker in international conferences in various universities. His research interests include IoT, artificial intelligence, employer branding, entrepreneurship and talent management.

Contributors

Balaji Dhanasekaran has completed his Ph.D. in Computer Science in the year 2010. He is having 22 years of overall teaching and administrative experience, including 17 years of abroad experience. He has guided more than 50 M.Phil., Masters and Undergraduate students' projects and research works. He is presently working as Head of the Department of IT in University of Technology and Applied Sciences, Salalah, Oman. He has published 14 journal papers, 34 international conference papers and 20 national conference papers. He has published few of his research work as book chapters. He has visited eight countries for presenting his research work. His research interests are IoT, ICT and M-Learning. He is one of the technical paper reviewers for many journals and international conferences. He has conducted various workshops for college students in India and Oman related to IoT, Research paper writing, Cloud Computing, etc. He is at present working on two research projects funded by the University of Technology and Applied Sciences, Salalah. His three research papers have got accepted by international conferences on Oman Vision 2040, organized by the Sultan Qaboos University, Oman.

Minisha Gupta is a researcher and trainer at Quality Cognition Private Limited, Delhi, India. She is currently working as a mentor and trainer for engineering and management students providing online lectures and workshops for them. She has more than 5 years of teaching and 7 years of post Ph.D. research experience in the field of Innovation management, Entrepreneurship, Leadership and Organizational change. Her research work focuses on innovation management, talent management, Green HRM, digital innovation and entrepreneurship. She has published many research articles in various national and international journals and presented research papers on national and international conferences of repute.

V. Harish has more than 17 years of industry and academic experience. He has served in various capacities like zonal manager, project management trainer, consultant. etc. A certified PMP (Project Management Professional) from Project Management Institute U.S.A., he has trained numerous professionals from industries such as IT, construction and banking. Also, as a consultant for lean practices for companies, he has helped many factories become more efficient. He has published a book titled *Practice Exercises for PMP®*. His area of interest is operations management, futuristic technologies, project management and lean manufacturing.

Haru Hong Khanh is associated with Saigon University as an Assistant Professor with more than three years of experience in academics. She is currently a lecturer in the faculty of Computer Science and Software Engineering, Saigon University, Vietnam. She is also short-term lecturer in some faculty of Software Engineering and Cryptocurrency Technology in private universities and institutes. She has supervised more than 15 undergraduate student projects. Her area of expertise includes finance and accounting, fintech, banking, blockchain and cryptocurrency technology, software engineering, database technology, and data engineering. She has taught courses such as Blockchain Technology, Software Engineering, and Data Engineering to graduate-level students.

D. Krishnaveni is a member of the faculty team at PSG Institute of Management, Coimbatore in the specialization of Human Resources. She is an alumnus of BITS, Pilani where she graduated in the dual disciplines of Biological Sciences and Electrical Engineering. After a few years in the software industry, where she worked on a variety of technologies and with clients across different continents, she moved to academics. Her academic efforts secured her a host of awards and recognitions. Along with her MBA, she also completed a Post Graduate Diploma in labour law and administrative law. Post her MBA, she pursued her Ph.D. in the area of sustainable agriculture. She is an avid researcher and has published a number of papers in reputed journals and has participated in international conferences. With a background in engineering, computing and the natural sciences, Krishnaveni believes she is uniquely poised to understand and relate to the sector-specific and commercial needs of businesses.

Neetu Kumari holds Ph.D. in Commerce, M.Phil in Commerce, M.Com., MBA (Human Resource Management and NET in Commerce). Dr Neetu Kumari has rich experience in teaching and research. She is currently working in the Department of Commerce, Udaipur Campus, University of Jammu, India. Her research interests focus on service marketing, entrepreneurship and human resource management with emphasis on consumer satisfaction, loyalty, service quality and perceived value. Her research work has been acknowledged in referred international journals like *Journal of Relationship Marketing*, *International Journal of Pharmaceutical and Healthcare Marketing*, *Management Research Review*, *Health Marketing Quarterly*, *Journal of Indian Business Research* and national journals like *NICE Journal* and *International Journal of Customer Relationship Management*, etc. She has authored books on E-commerce, digital marketing and women entrepreneurship which was enjoyed tremendous success and was highly appreciated by academicians, scholars and students. She is a frequent speaker in national and international conferences on several topics related to marketing, human resource and entrepreneurship. She has authored more than 30 research papers, case studies and chapters in books including papers listed in Scoupus and ABDC ranked journals. She has also received best paper presentation award in national conference (2021) organized by Social Science & Management Welfare Association Jammu & Kashmir, Youth Economic Association J&K.

A. Mansurali is an academician in the field of business management, with special focus on marketing, analytics and applied research. With innovative and engaging pedagogy, he engages and trains students to understand the current developments in the field and has several academic contributions in the form of publications to his credit. He is proficient in R, Python, Tableau, Power BI and machine learning algorithms. He is also a certified trainer for R and Machine learning from STAR certification. His competence in analytics and research has earned him key roles in

Contributors

sponsored research projects. He has spent ten years in teaching management graduates marketing, analytics related courses and researching the same. He has also bagged funding from the University Grants Commission to conduct research in the area of Microfinance.

Yogesh Pant is an assistant professor at Himalayan School of Science and Technology in Swami Rama Himalayan University, India. He is member of international association of engineers (IAENG). He has been teaching since 2014 in different disciplines of computer science and artificial intelligence. His research focuses primarily on machine learning, Internet of Things and quantum computing. He has presented many of his work in many international conferences and authored many book chapters in cloud computing, IoT and artificial intelligence application in business.

Bhakti Parashar is an assistant professor of Economics at Vellore Institute of Technology-Bhopal University. She is at present program chair of VIT-Business School as well. She completed her Ph.D. at Barkatulla University, Bhopal, India. She has been teaching since 2007 in various disciplines of economics and management such as managerial economics, environmental studies, engineering economics & business organization, microeconimics, macroeconimics, talent management and emotional intelligence. Her research focus is mainly on Economics, Management and Social Sciences. She has published articles on health economics, education, demography, conflict management, employer's branding and globalization, digital marketing in journals, book chapter conference proceedings. She has international book chapter publications in NOVA publication USA and presented research papers in national and international conferences as well.

S.P. Preethi is a student pursuing a masters degree in the Department of Information Technology (M.Tech), Madras Institute of Technology, Anna University, Chennai. Her research focuses primarily on machine learning, deep learning and remote sensing.

R. Shanmuga Priya is presently working in the Department of Information Technology, Madras Institute of Technology Campus, Anna University, Chennai. She is currently pursuing her Ph.D. in College of Engineering, Guindy, Anna University. Her research focuses primarily on image processing, remote sensing, deep learning, video processing analytics and artificial intelligence. She has published research articles and presented research papers on international conferences in image processing and machine learning. She has also authored books in big data analytics and artificial intelligence.

X. Mercilin Raajini received Ph.D. degree in the faculty of Electrical Engineering, Anna University, Chennai. She is currently working as an associate professor in Prince Shri Venkateshwara Padmavathy Engineering College, Ponmar, Chennai, India. She published more than 20 papers in conferences and international journals. She authored three books in the area of wireless sensor network. She has 13 years of teaching experience. She is a life member in The Indian Society for Technical Education. She guided more than 30 under graduate and post graduate students. Her area of interest includes wireless sensor networks, signal processing, IOT applications, drug screening and computational optimization.

L. Rajesh is presently working in the Department of Electronics Engineering, Madras Institute of Technology Campus, Anna University, Chennai, Tamil Nadu, India. He obtained his Ph.D. in wireless networks and Master of engineering in communication and networking from Anna University, Chennai, Tamil Nadu, India. His area of interest includes wireless networks, game theory. He has number of international and national journal publications.

G. Rajesh working as an assistant professor, Department of Information Technology of Anna University, Chennai, India. He completed his Ph.D. from Anna University, Chennai in wireless sensor networks. He has around 12 years of teaching and research experience. His area of research interest includes wireless sensor networks and its IoT applications, software engineering and computational optimization. He published more than 20 research papers in journals and conferences.

Ashwini Kumar Saini is an assistant professor at the Faculty of Computer Science and Engineering, G. B. Pant Institute of Engineering and Technology, Pauri Garhwal. He has 11 years of teaching experience. He has published many research papers on different topics on emerging technology in national and international journals. He has participated and presented research papers on national and international conferences on various topics in India. His area of interest is educational data mining, machine learning, IoT.

Mukhtar Sama is an assistant professor at Mechanical Engineering Department Marwadi University, Rajkot. He holds Degree in Masters of Engineering in Production Engineering from MS University Baroda. He has more than nine years of experience in academics. He has been teaching different subject like, manufacturing processes, operation research, data mining & analysis and IoT in manufacturing. His research primarily focuses on implementation of IoT in manufacturing processes.

1 The Role of Artificial Intelligence in Adopting Green HRM Practices

Minisha Gupta

CONTENTS

1.1 Introduction 1
1.2 Review of Literature 2
    1.2.1 Artificial Intelligence 2
    1.2.2 Green Human Resource Management 4
    1.2.3 Determinants of GHRM Adoption 5
1.3 Role of AI in Adopting GHRM Practices 8
1.4 Conclusion 11
References 12

1.1 INTRODUCTION

Artificial intelligence (AI) is constantly trying to meet the criteria of Industry 4.0 by transforming traditional organizations into smart factories where human efforts can be minimized and their talent can be leveraged for attaining organizational sustainability (Kshetri, 2021). However, business organizations in developing economies are struggling both internally and externally. On one hand, while organizations have to meet the demands of Industry 4.0 by transforming themselves into smart factories, they on the other hand also have to be responsive to the changing expectations of their customers and environment. With the increasing pressures of maintaining the environment, reducing waste and implementing cleaner production policies, organizations are shifting their focus to implement green human resource management (green HRM) practices (Pham et al., 2019). In order to both meet the ends and satisfy the organizational stakeholders, both internally and externally, companies are relying on AI (Garg et al., 2018). AI supports organizations with the advanced digital technologies, cloud computing and data storage facilities, decision-making applications and smart analytical tools (Kshetri, 2021). This chapter will explain the role of AI in adopting and implementing green HRM practices. The next section will discuss the literature carried out on green HRM and artificial intelligence. The latest examples and cases from the industry will be discussed in the chapter.

DOI: 10.1201/9781003145011-1

1.2 REVIEW OF LITERATURE

1.2.1 ARTIFICIAL INTELLIGENCE

AI is a not a new concept anymore for developing or developed economies. AI is the science and engineering of making intelligent machines, especially intelligent computer programs (McCarthy, 1989). AI is an innovative tool used for transforming the role of management and organizational practices (Kshetri, 2021; Sharma & Rana, 2021). It is a subpart of computer science, which is concerned with making computers sophisticated so that they can act smartly (Nilson, 1980). It includes various intelligent tasks performed by computers on behalf of human beings such as recognition, reasoning and learning (Hilker, 1986). AI has changed the pattern of work and decision-making abilities for many organizations with its smart technological approach such as genetic algorithms, neural networks, data mining, text mining, sentiment analysis and interactive voice recognition applications (Lauterbach, 2019; Strohmeier & Piazza, 2015). It improves the decision-making ability and cost-effectiveness of the organizations by making decisions on real-time data (Kahraman et al., 2011; Rana & Sharma, 2017; Lawler & Elliot, 1996). However, the question arises: What is artificial intelligence? Is it a tool, an application, software, a methodology or a thought? AI cannot be defined in terms of a written definition, but various researchers defined artificial intelligence in technical and sociological terms.

A review study tries to explain artificial intelligence through the lens of social sciences and technology (Lauterbach, 2019). As per social sciences, AI has been described as artificial narrow intelligence, artificial general intelligence and artificial super intelligence. *Artificial narrow intelligence* (ANI) focuses on specializing one task or one function of the organization, which might be scheduling a meeting, recognizing voices, or human touch. Statistical machine learning in games (chess, Go and Jeopardy!), self-driving cars, automated assistants (Apple Siri, Amazon Alexa, Google Now and Microsoft Cortana) are few examples of ANI (Bundy, 2017). *Artificial general intelligence* (AGI) is interdisciplinary in nature and provides multiple solutions for single problems which help the decision-makers to select the appropriate solution out of the available ones (Voss, 2007). AGI applications use sensory data (visual or auditory) and deal with attributes to present simplified data for researchers and analysts. *Artificial super intelligence* (ASI) is defined as, "an intellect that is much smaller than the best human brains in practically every field, including scientific creativity, general wisdom and social skills" (Bostrom, 2014, pp. 109–114).

On the other hand, AI has also been explained through technological background. Technological AI is further disseminated in machine learning technologies and perception technologies. *Machine learning technologies* (MLT) focus on copying the human skills and thinking to presume in order to provide better analytical results. It helps in improving the efficiency and performance automatically for a given task by observing the relevant data. It is mostly used for research and development projects, speech recognition, lie detection, image recognition and several other tasks which depend on human intelligence and decision-making skills. *Perception technologies* include computer vision and natural language processing. *Computer vision* is a form of machine learning which includes recognition of faces, developing predictions, collecting and tagging data, assessing and analyzing data for manufacturing and service industries. *Natural language processing* is a form of AI whereby individuals can speak with machines and get their work done; for example, the virtual personal assistant Alexa (Amazon Echo), and call center agents.

The form of artificial intelligence defined here is shown in Figure 1.1.

Figure 1.1 explains the different dimensions of AI which are adopted and implemented by organizations to improve their processes and services. These AI technologies are not only changing human effort, but also generating more opportunities to leverage the human skills. AI is not limited to manufacturing or service organizations but is also implemented for improving organizational efficiency and effectiveness. Companies are employing AI tools and techniques for various administrative and HRM functions including recruitment, training and development, performance appraisal, career development and talent retention.

Modern management theories management by objective (MBO) and management by exception (MBE) are focusing more on green environment issues (Chan & Chan, 2004; Chan et al., 2014; Mařík & Lažanský, 2007). Therefore, along with the challenge to sustain in this competitive scenario, companies also have to maintain green HRM practices and functions. With smart and innovative applications, AI helps the organizations attain green HRM functions and practices (Sekhri & Cheema, 2019). In the era of Industry 4.0 when organizations are transforming into digitalized systems, AI plays a great role in turning imagination into reality. With its smart and digitally equipped applications, AI is of great help for organizations

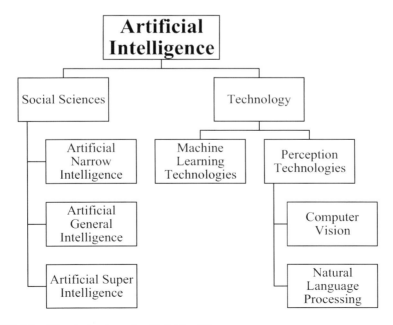

FIGURE 1.1 Discrimination of artificial intelligence.

to adopt and implement green HRM practices and functions. Applications like chat boats, digital attendance, job intelligence maestros, automation, distance assistants and e-pass systems are the major AI tools which have been employed by organizations to improve their HRM functions and processes. Similarly, as per the scenario AI tools and applications are used by organizations to adopt green HRM (GHRM) functions and processes. The next section will discuss the main GHRM functions practiced in organizations.

1.2.2 Green Human Resource Management

With the constantly increasing pressure of environmentalists to maintain and preserve a pollution-free environment, organizations cannot avoid the implementation of GHRM practices (Jabbour, 2013; Pham et al., 2019). Another reason for the same is increasing customer awareness and concern of job seekers about the organizational policies toward maintaining a sustainable and green environment (Teixeira et al., 2012). Organizations cannot attain sustainable competitive advantage without leveraging the talent of their employees, and talented employees can only be retained with effective HRM policies. Therefore, GHRM has attracted the interest of various scholars, researchers, policymakers and academics after the research initiated by a small group of researchers (Jabbour & Santos, 2008; Tomer & Rana, 2020; Jackson et al., 2011; Pham et al., 2019; Renwick et al., 2016; Yong et al., 2019a, 2019b). Implementation of GHRM practices also helped organizations to minimize carbon emissions, go for paperless approaches and deduct waste production in manufacturing units (Ahmad, 2015; Yong et al., 2019a). GHRM has been considered as a sustainable change for the organizations (Sawang & Kivits, 2014).

GHRM can be defined as HRM-related aspects of environmental management (EM) and focuses on the role of HRM in preventing environmental pollution using organizational procedures (Renwick et al., 2008; Renwick et al., 2013). GHRM practices have been considered a combination of original HRM practices and strategic HRM measures of organizations to preserve and sustain green environment (Gholami et al., 2016). GHRM practices depend on three components: developing green ability, motivating green employees and facilitating green opportunity to the employees (Renwick et al., 2013). These practices and measures revolve around the main HRM functions which are disseminated into recruitment and selection, training and development, performance management, compensation and reward management, and finally talent retention. HRM functions support organizations in implementing environmental measures by formulating environmental goals (Cohen, 2012). HRM also serves as a partner to formulate environment-based corporate values and design strategies to attain sustainability. Therefore, HRM functions are further divided on the basis of antecedents of GHRM practices. Table 1.1 divides the HRM functions as per the factors of GHRM practices.

These identified GHRM functions have been studied by researchers, leading to an increase in GHRM studies from 2016 (Cabral & Dhar, 2019; Chaudhary, 2018, 2019a, 2019b; Dumont et al., 2017; Guerci et al., 2016; Jabbour & Renwick, 2018; Pham et al., 2019; Shah, 2019; Srivastava et al., 2020; Yong et al., 2019a; Yusliza et al., 2017). These studies explain concepts and models of GHRM, implementation

The Role of AI in Adopting Green HRM

TABLE 1.1
HRM Functions Disseminated on the Basis of GHRM Practices

No	Antecedents of GHRM Practices	HRM Functions	Research Studies
1	Developing green ability	• Recruitment and selection • Training and development • Job description	Jabbour et al. (2010); Jabbour (2013); Renwick et al. (2013); Pham and Paillé (2019); Ren et al. (2018); Zaid et al. (2018); Chaudhary (2018); Chaudhary (2019a); Roscoe et al. (2019)
2	Motivating green employees	• Performance management/appraisal • Reward and pay system	Masri and Jaaron (2017); Gupta (2018); Longoni et al. (2018); Aboramadan (2020)
3	Facilitating green opportunities	• Employee involvement and empowerment • Supportive climate/culture • Union's role in EM • Organizational learning	Jabbour and Santos (2008); Teixeira et al. (2012); Pinzone et al. (2016); Gholami et al. (2016); Luu (2017); Masri and Jaaron (2017); Tang et al. (2018); Gupta (2018); Ren et al. (2018); Chaudhary (2019a); Kim et al. (2019); Fawehinmi et al. (2020); Aboramadan (2020); Liu et al. (2020); Rubel et al. (2021); Song et al. (2020); Srivastava et al. (2020)

Sources: From Pham et al. (2019) and Yong et al. (2019a)

of GHRM, determinants of GHRM adoption and outcomes of GHRM adoption (at individual and organizational levels). Since the objective of this study is to find the role of artificial intelligence in adopting GHRM thus, it will focus only on the determinants and outcomes of GHRM adoption which are discussed in the next section.

1.2.3 DETERMINANTS OF **GHRM** ADOPTION

With the increasing pressures of customers, stakeholders and orientation toward the environment, organizations have to look forward toward reducing production waste and promoting environmental sustainability by taking green initiatives through HRM practices (Guerci et al., 2016; Yong & Mohd-Yusoff, 2016). Employees' intellectual capital also significantly affects HRM practices (Kong & Thomson, 2009). An empirical study was conducted on 112 Malaysian manufacturing firms on the basis of intellectual capital–based view theory which revealed that green human capital and green relational capital significantly influence the organization's incentive to implement green HRM practices (Yong et al., 2019b). Green human capital consist of employees' knowledge, skills, capabilities, experience, attitudes, wisdom, creativity and commitment about green environment or green innovation (Chen, 2008; Li & Chang, 2010; Chahal & Bakshi, 2014). Organizations with such human capital find it easy to implement green HRM practices. Similarly, green relational capital is a combination of an organization's interactive relationship with customers, suppliers, network members and partners about corporate environmental management and green innovation, which enables it to create fortunes and obtain competitive advantage (Chen, 2008). As customers, suppliers, employees and partners are the main stakeholders of the organization, their impact on implementing green HRM services is relatively greater.

Moreover, green HRM practices can only be adopted and implemented in the organizations if top management provide support to these practices. Top management holds the decision to initiate any change in the organization, and they must have orientation toward green environment which helps them to support green HRM practices (Obeidat et al., 2020; Sawang & Kivits, 2014; Teixeira et al., 2012). With effective leadership and employee commitment, top management initiates changes to the organizational culture by adopting green HRM practices such as green recruitment and selection, green job descriptions, green training and development, green performance appraisal system, green safety and health provision, green employee and labor relations, and green grievance handling systems (Ahmad, 2015; Chaudhary, 2019b; Gholami et al., 2016; Guerci et al., 2016; Jabbour, 2013; Jabbour et al., 2010; Jackson et al., 2011; O'Donohue & Torugsa, 2016; Ren et al., 2018; Rana & Sharma, 2019; Renwick et al., 2013; Shah, 2019; Siyambalapitiya et al., 2018; Srivastava et al., 2020; Tang et al., 2017; Zibarras & Coan, 2015). Adoption of green HRM practices help organizations attain green environmental goals by transforming employee behavior and attitudes toward adopting green behaviors and developing a green workforce (Aboramadan, 2020; Cabral & Dhar, 2019; Cabral & Dhar, 2020; Chaudhary, 2019b; Dumont et al., 2017; Fawehinmi et al., 2020; Islam et al., 2020; Jackson & Seo, 2010; Kim et al., 2019; Luu, 2017; Luu, 2018; Mishra, 2017; Aboramadan, 2020; Mukherjee & Chandra, 2018; Ojo et al., 2020; Renwick et al., 2013; Rubel et al., 2021).

Green recruitment and selection helps organizations attract more talented employees due to their policies as most of the Generation Y employees look for employers with green HRM practices (Chaudhary, 2018). Employers have transformed themselves as green employer to attract talent (Ahmed, 2015). European companies like Siemens, BASF, Bayer, Mannesmann and Rover Group use environmental activities as part of their job descriptions (Wehrmeyer, 1996). A systematic review was carried out on 22 peer-reviewed articles and identified the recommendations of recruitment and selection, along with mediators and moderators, to establish a relationship between applicant attraction outcomes and corporate environmental sustainability (Pham & Paillé, 2019).

In order to transform the employees into a green workforce, organizations are integrating environmental performance with performance management systems to prevent any damage to the environment (Epstein & Roy, 1997). Companies install environmental performance standards and green information systems to attain useful data on employees' organizational performance (Alfred & Adam, 2009). They also initiate green training and development to educate employees about managing environment, minimizing waste, water preservation, conserving energy and providing them with opportunities to identify sustainable innovations to preserve the environment (Zoogah, 2011). Previous research studies have unraveled the determinants of employee green behavior as it is important for implementing green HRM policies (Aboramadan, 2020; Islam et al., 2020; Liu et al., 2020; Ojo et al., 2020; Rubel et al., 2021). Green training and development is an important construct which affect employees' behavior toward environmental safety and prevention. It also makes employees aware of the different aspects of environmental management as they look for more eco-friendly options like paperless documentation, energy consumption, waste reduction, recycling scraps and generating creative and innovative solutions. Training in U.S. firms involves regulatory requirements, employees' awareness of environment management and training on environmental quality management (TEQM) (Milliman & Clair, 1996).

Apart from these, companies are also introducing green compensation policies whereby employees are rewarded and encouraged to adopt eco-friendly activities. More than 8% of firms in the United Kingdom are rewarding their employees' green behaviors with financial or non-financial incentives (Phillips, 2007). Employees' green behavior and their participation in environmental activities is also considered and awarded by companies as their commitment toward environment management (Forman & Jorgensen, 2001). To maintain a healthy employee–employer relations, companies are also adopting green safety and health policies, green employee/labor relations and green grievance handling systems. These policies facilitate employee participation and empowerment toward environment management activities. Previous studies found that involving and empowering employees result in reduced waste and pollution, along with efficient use of resources (Florida & Davison, 2001; Kitazawa & Sarkis, 2000; Renwick et al., 2008; Wee & Quazi, 2005). Figure 1.2 describes the determinants of GHRM practices in organizations.

FIGURE 1.2 Determinants of adopting green HRM practices.

Source: From Yong et al. (2019a)

Figure 1.2 shows the various factors to which organizations have to adopt GHRM practices. In a green HR survey, it was revealed that more than 54% of companies have infused green management practices in their organizational activities, 74% used digital innovation to conduct meetings without traveling, 76% of organizations promoted paperless documentation, 60% have adopted employee fitness and wellness programs, and 80% used artificial intelligence to perform their activities (Aggarwal & Sharma, 2015). With the smart and artificially intelligent software, adoption and implementation of GHRM services becomes easy for organizations (Garg et al., 2018). For instance, *PepsiCo* has implemented AI software to interview and select candidates for their vacant positions. The software automatically scan CVs (curriculum vitae) from various job portals and selects the candidates having matching profiles, which minimizes the job of interviewers to travel across countries to interview the candidates. Infosys implemented new ERP system and smart meters and able to reduce their energy consumption by 85%. Companies like ITC, Lufthansa Group, Hyatt Group of Hotels, Nokia, Gensol Consultants Private Limited and id8 Media Solutions have also adopted AI technologies to implement GHRM practices for adopting green and healthy environment. The next section will discuss about the role of AI in adopting GHRM practices.

1.3 ROLE OF AI IN ADOPTING GHRM PRACTICES

Organizations—whether in developing or developed economies—have already adopted various modern technologies to improve their operations and processes. For instance, using Big Data analytics to analyze the trends and patterns of recruitment, sales and production has now become common for organizations, but implementing AI for adopting GHRM practices is still in its nascent stage. With its smart and solution oriented technologies, AI has become the key to attain sustainable competitive advantage; however, it is still limited to only few companies. One major reason for this reluctance is the inability of the talent to manage such applications and another is the fear of losing jobs as machines replace human effort. On contrary, AI improves work efficiency and reduces human efforts which can be rather used for attaining green practices. Recently, an application has been developed by an Indian IT company with the name "AskDexter" which manages 22,000 employees by handling HR functions like solving employees' queries about leave and company policies (Garg et al., 2018). This app also provides information about upcoming job opportunities and openings. It also helps job seekers to align their skills with the job requirements. AI is helpful to perform various HR functions. Figure 1.3 explains the role of AI in GHRM practices.

Figure 1.3 explains that AI supports organizations in developing green ability of its employees by providing applications which are helpful in recruitment and selection, training and development, and identifying candidates as per the job description. AI applications screen and select the CVs of job seekers and select the best ones matching the requirement of the organization. AI applications have reduced recruiters' jobs up to 60% through automated and digitalized technologies (Stahl, 2021). AI will reduce 80% of recruiters' jobs through HR help desk applications in the future. Moreover, AI applications also help in reducing cost of traveling for interviewers and

The Role of AI in Adopting Green HRM

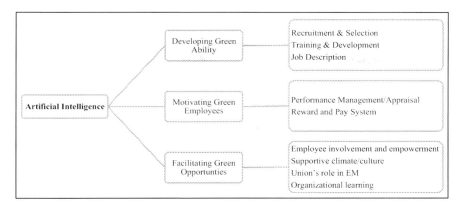

FIGURE 1.3 Role of AI in adopting GHRM practices.

recruiters as they can take interviews using portals like Zoom and Microsoft Teams. Companies also use AI applications to train their employees like Google Classroom, or Google Meet platforms to train their employees online.

AI applications also help organizations in motivating employees to go green by using performance appraisal software and data analytics. Companies assess employees' performance on daily, quarterly, and yearly bases, and provide them with feedback for self-appraisal and help in overcoming their weaknesses. This feedback also helps employees to identify how they can reduce waste, what practices they can follow to maintain a healthy lifestyle and how to focus on creative solutions for maintaining a green environment. Companies monitor their employees' activities using AI applications and reward the employees following green practices. Organizations find it challenging to meet the standards of green performance management and appraisal as it goes through various departments and staff units. Companies are trying to understand these issues and they are trying to upgrade their performance appraisal systems by initiating green information systems to manage the standards of green performance management.

Various AI applications like chatbots and chat rooms help employees solve their queries about leave and organization systems, which saves a lot of time for counselors and mentors which can be used in other productive activities. These applications are not only employee friendly but also help in improving employee–employer relationships, solving grievances and developing green corporate culture. Companies manage their operations like manufacturing processes, warehouse management, and supply chain management using AI applications which are smart and help reduce waste and save energy and resources. With the use of AI applications and technologies, the chance of error reduces and companies can easily adopt GHRM practices.

Automation through AI is taking place in GHRM services which are depicted from the AMO (ability, motivation, opportunity) framework shown in Figure 1.4.

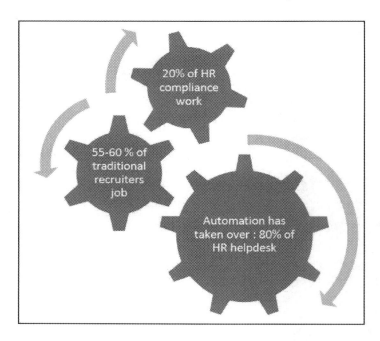

FIGURE 1.4 AMO (ability, motivation, opportunity) framework in HRM.
Source: Garg et al. (2018)

TABLE 1.2
AI Technologies for Adopting GHRM Practices

No.	Technology	Company	Outcome
1	Trackerapp	Pegasystems	This app is designed to help organizations track the spread of COVID-19 among its employees.
2	Virtual training	Samsung	Providing virtual or e-internships to new recruits of Generation Z category to go through Team Samsung Online portal.
3	Jobs Intelligence Maestro	DBS	Through JIM, talent acquisition specialists take only eight minutes to assess a candidate's résumé, as compared to the typical 30 minutes.
4	Automation	EPFO	By launching its first fully automatic claim settlement system in a record time of just five days, EPFO disbursed a whopping Rs 11,540 crore to its members in April–May 2020.
5	Distance Assistant	Amazon	Distance Assistant is a new technology driven by machine learning (ML) and augmented reality (AR) which uses depth sensors to provide visual cues to workers, to maintain safe distance from each other.
6	LTI Safe Radius	LTI (Larsen & Turbo Infotech)	This app has been launched to make sure that the health of the workforce remains protected as they return to office.
7	Chatbot Nina	Sun Life ASC India	This chatbot answers employees' queries, and around 74% of users have reported being satisfied with responses to their queries.

The Role of AI in Adopting Green HRM

No.	Technology	Company	Outcome
8	Digital HRM system	Telangana State Police	The automated HRMS will efficiently manage all aspects, including recruitment, employee data (service register), time and attendance, absence and leave, performance and rewards, training, learning and development, administration, user analytics and personnel grievances.
9	E-Pass system	Indian Western Railway	It is the paperless version of the privilege passes or privilege ticket given to employees of the Indian Railways and their families.
10	Creating talent cloud	Matrimony.com	The app creates talent cloud to select the experienced employees and use real-time data to improve employee productivity.
11	Talent acquisition	Microsoft Teams	This allows talent experience management (TXM) for Teams to come together to reduce collaborative attrition for many small-scale and large-scale enterprises for communication.
12	HRMS	Indian Railways	This system will improve productivity and ensure employee satisfaction.
13	HR Tools	Microsoft	VIVA has been launched for better employee experience in terms of communications knowledge, learning, resources and insights, along with consulting and advisory services.

Sources: Data accessed from HRKatha.com, PeopleMatters.com

Companies are using automation like data mining techniques to select and retain green talent. Green employees have the ability to perform tasks as per environmental requirements; they can be motivated to maintain the environment and given the empowerment to produce creative solutions for sustaining the environment. Companies use AI tools to enhance learning and development among their employees like free digital libraries to explore content. There are few other AI techniques implemented for adopting sustainable GHRM practices, as shown in Table 1.2.

1.4 CONCLUSION

In this dynamic scenario, when companies are shifting their priorities on attaining sustainable competitive advantage rather than achieving profits, then it also becomes their obligation to preserve the nature and environment. It will also make organizations the preferable employer brands for the job seekers, and they will be able to attract and retain talent. Thus, organizations have to take measures for implementing green HRM policies. Along with attracting and retaining talent, companies also have to take care of their existing workforce, because any random change can create resistance among the employees (Gupta & Haque, 2015; Gupta & Lenka, 2018). There comes the role of AI, which helps organizations to adopt GHRM services by offering various applications and smart technologies which simplify the work and increase productivity of employees. AI is also important to be undertaken by organizations because of diverse workforce which includes Baby Boomers (born 1946–1960), Generation X (born 1961–1980), Generation Y (born 1981–2000) and Generation Z (born 1995–2012) (Naim & Lenka, 2017; Singh, 2014). Gen Y is achievement oriented and motivated by growth needs; therefore, their motivation to adopt GHRM practices is different (Wong et al., 2008). A study identified that Generation Y employees take care of the planet and they expect to be rewarded for adapting GHRM practices (Jain & D'lima, 2018). On the contrary, Generation Z is tech-savvy, empowered, more realistic and pre-matured; therefore, they can be motivated by using technology-driven tools and techniques (Singh, 2014; Sharma et al., 2021). So, with diversified workforces, organizations have to use AI tools and techniques to adopt GHRM practices. Previous studies identified training, recruitment and rewards as the important factors for implementing GHRM practices (Nobari et al., 2018). Thus, AI tools and techniques can be used for these processes to engage and motivate employees to voluntarily participate in green activities. Adoption of green HRM activities is beneficial for environmental performance and organizational sustainability (Obeidat et al., 2020).

The present study unravels the role of AI technologies and applications in adopting green HRM practices. The determinants of GHRM practices have been explained in detail, which helps in identifying the relevant AI applications for the same. Previous studies also suggested that for effective implementation of e-HRM practices, use of AI, Big Data analytics and people/HR analytics should be taken into consideration (Poba-Nzaou et al., 2020; Prakash et al., 2021). However, this study is based on review of literature only, but can be explored further through case study and empirical research work. Future studies can be conducted to identify the role of AI in adopting GHRM practices and to explore the niche areas of GHRM research, and can also be researched through conceptual or empirical work. Future research work can also be done to identify the motivating factors at individual, group/team and organizational levels through which employees can be influenced to voluntarily pursue GHRM practices. Future studies can also take into account the relationship between GHRM and green innovation as a step toward attaining sustainability.

REFERENCES

Aboramadan, M. (2020). The effect of green HRM on employee green behaviors in higher education: The mediating mechanism of green work engagement. *International Journal of Organizational Analysis*. https://doi.org/10.1108/IJOA-05-2020-2190

Aggarwal, S., & Sharma, B. (2015). Green HRM: Need of the hour. *International Journal of Management and Social Science Research Review*, *1*(8), 63–70.

Ahmad, S. (2015). Green human resource management: Policies and practices. *Cogent Business & Management*, *2*(1), 1030817. https://doi.org/10.1080/23311975.2015.1030817

Alfred, A. M., & Adam, R. F. (2009). Green management matters regardless. *Academy of Management Perspectives*, *23*(3), 17–26.

Bostrom, N. (2014). *Superintelligence: Paths, dangers, strategies*. Oxford: Oxford University Press.

Bundy, A. (2017). Review of preparing for the future of artificial intelligence. *AI and Society*, *32*(2), 285–287. https://doi.org/10.1007/s00146-016-0685-0

Cabral, C., & Dhar, R. L. (2019). Green competencies: Construct development and measurement validation. *Journal of Cleaner Production*, *235*, 887–900.

Cabral, C., & Dhar, R. L. (2020). Green competencies: Insights and recommendations from a systematic literature review. *Benchmarking: An International Journal*, *28*(1), 66–105.

The Role of AI in Adopting Green HRM

Chahal, H., & Bakshi, P. (2014). Effect of intellectual capital on competitive advantage and business performance: Role of innovation and learning culture. *International Journal of Learning and Intellectual Capital, 11*(1), 52–70.

Chan, E. S., Hon, A. H., Chan, W., & Okumus, F. (2014). What drives employees' intentions to implement green practices in hotels? The role of knowledge, awareness, concern and ecological behaviour. *International Journal of Hospitality Management, 40*, 20–28.

Chan, F. T., & Chan, H. K. (2004). A comprehensive survey and future trend of simulation study on FMS scheduling. *Journal of Intelligent Manufacturing, 15*(1), 87–102.

Chaudhary, R. (2018). Can green human resource management attract young talent? An empirical analysis. *Evidence-Based HRM, 6*(3), 305–319.

Chaudhary, R. (2019a). Green human resource management and job pursuit intention: Examining the underlying processes. *Corporate Social Responsibility and Environmental Management, 26*(4), 929–937.

Chaudhary, R. (2019b). Green human resource management and employee green behavior: An empirical analysis. *Corporate Social Responsibility and Environmental Management.* https://doi.org/10.1002/csr.1827

Chen, Y. S. (2008). The driver of green innovation and green image: Green core competence. *Journal of Business Ethics, 81*(3), 531–543.

Cohen, D. J. (2012). Identifying the value of HR certification: Clarification and more complex models required. *Human Resource Management Review, 22*(4), 258–265.

Dumont, J., Shen, J., & Deng, X. (2017). Effects of green HRM practices on employee workplace green behavior: The role of psychological green climate and employee green values. *Human resource management, 56*(4), 613–627.

Epstein, M., & Roy, M. (1997). Using ISO 14000 for improved organizational learning and environmental management. *Environmental Quality Management, 7*, 21–30.

Fawehinmi, O., Yusliza, M. Y., Mohamad, Z., Faezah, J. N., & Muhammad, Z. (2020). Assessing the green behaviour of academics: The role of green human resource management and environmental knowledge. *International Journal of Manpower, 41*(7), 879–900.

Florida, R., & Davison, D. (2001). Gaining from green management: Environmental management systems inside and outside the factory. *California Management Review, 43*(3), 64–84.

Forman, M., & Jorgensen, M. S. G. (2001). The social shaping of the participation of employees in environmental work within enterprises-experiences from a Danish context. *Technology Analysis & Strategic Management, 13*(1), 71–90.

Garg, V., Srivastav, S., & Gupta, A. (2018, October). Application of artificial intelligence for sustaining green human resource management. In *2018 International Conference on Automation and Computational Engineering (ICACE)* (pp. 113–116). IEEE, Noida, India.

Gholami, H., Rezaei, G., Saman, M. Z. M., Sharif, S., & Zakuan, N. (2016). State-of-the-art Green HRM System: Sustainability in the sports center in Malaysia using a multi-methods approach and opportunities for future research. *Journal of Cleaner Production, 124*, 142–163.

Guerci, M., Longoni, A., & Luzzini, D. (2016). Translating stakeholder pressures into environmental performance: The mediating role of green HRM practices. *The International Journal of Human Resource Management, 27*(2), 262–289.

Gupta, H. (2018). Assessing organizations performance on the basis of GHRM practices using BWM and Fuzzy TOPSIS. *Journal of Environmental Management, 226*, 201–216.

Gupta, M., & Haque, M. M. (2015). Talent retention: A major concern for organizations. In *Proceedings of ICRBS, 2015*, IIT Roorkee, Greater Noida, India.

Gupta, M., & Lenka, U. (2018). Employer branding: A talent retention strategy using social media. *Pragyaan: Journal of Mass Communication*, 21.

Hilker, E. (1986). Artificial intelligence: A review of current information sources. *Collection Building*, *7*(3), 14–30.

Islam, M. A., Jantan, A. H., Yusoff, Y. M., Chong, C. W., & Hossain, M. S. (2020). Green Human Resource Management (GHRM) practices and millennial employees' turnover intentions in tourism industry in malaysia: Moderating role of work environment. *Global Business Review*, *122*(1), 1–21. https://doi.org/10.1177/0972150920907000.

Jabbour, C. J. C. (2013). Environmental training in organisations: From a literature review to a framework for future research. *Resources, Conservation and Recycling*, *74*, 144–155.

Jabbour, C. J. C., & Renwick, D. W. S. (2018). The soft side of environmentally-sustainable organizations. *RAUSP Management Journal*, *53*(4), 622–627.

Jabbour, C. J. C., & Santos, F. C. A. (2008). Relationships between human resource dimensions and environmental management in companies: Proposal of a model. *Journal of Cleaner Production*, *16*(1), 51–58.

Jabbour, C. J. C., Santos, F. C. A., & Nagano, M. S. (2010). Contributions of HRM throughout the stages of environmental management: Methodological triangulation applied to companies in Brazil. *The International Journal of Human Resource Management*, *21*(7), 1049–1089.

Jackson, S. E., & Seo, J. (2010). The greening of strategic HRM scholarship. *Organization Management Journal*, *7*(4), 278–290.

Jackson, S. E., Renwick, D. W., Jabbour, C. J., & Muller-Camen, M. (2011). State-of-the-art and future directions for green human resource management: Introduction to the special issue. *German Journal of Human Resource Management*, *25*(2), 99–116.

Jain, N., & D'lima, C. (2018). Green HRM: A study on the perception of Generation Y as prospective internal customers. *International Journal of Business Excellence*, *15*(2), 199–208.

Kahraman, C., Kaya, İ., & Çevikcan, E. (2011). Intelligence decision systems in enterprise information management. *Journal of Enterprise Information Management*, *24*(4), 360–379.

Kim, Y. J., Kim, W. G., Choi, H. M., & Phetvaroon, K. (2019). The effect of green human resource management on hotel employees' eco-friendly behavior and environmental performance. *International Journal of Hospitality Management*, *76*, 83–93.

Kitazawa, S., & Sarkis, J. (2000). The relationship between ISO 14001 and continuous source reduction programs. *International Journal of Operations & Production Management*, *20*, 225–248.

Kong, E., & Thomson, S. B. (2009). An intellectual capital perspective of human resource strategies and practices. *Knowledge Management Research & Practice*, *7*(4), 356–364.

Kshetri, N. (2021). Evolving uses of artificial intelligence in human resource management in emerging economies in the global South: Some preliminary evidence. *Management Research Review*. https://doi.org/10.1108/MRR-03-2020-0168

Lauterbach, A. (2019). Artificial intelligence and policy: Quo vadis? *Digital Policy, Regulation and Governance*, *21*(3), 238–263.

Lawler, J. J., & Elliot, R. (1996). Artificial intelligence in HRM: An experimental study of an expert system. *Journal of Management*, *22*(1), 85–111.

Li, Q., & Chang, C. (2010). The customer lifetime value in Taiwanese credit card market. *African Journal of Business Management*, *4*(5), 702–709.

Liu, Z., Mei, S., & Guo, Y. (2020). Green human resource management, green organization identity and organizational citizenship behavior for the environment: The moderating effect of environmental values. *Chinese Management Studies*. https://doi.org/10.1108/CMS-10-2019-0366

The Role of AI in Adopting Green HRM

Longoni, A., Luzzini, D., & Guerci, M. (2018). Deploying environmental management across functions: The relationship between green human resource management and green supply chain management. *Journal of Business Ethics, 151*(4), 1081–1095.

Luu, T. T. (2017). CSR and organizational citizenship behavior for the environment in hotel industry. *International Journal of Contemporary Hospitality Management, 29*(11), 2867–2900.

Luu, T. T. (2018). Employees' green recovery performance: The roles of green HR practices and serving culture. *Journal of Sustainable Tourism, 26*(8), 1308–1324.

Mařík, V., & Lažanský, J. (2007). Industrial applications of agent technologies. *Control Engineering Practice, 15*(11), 1364–1380.

Masri, H. A., & Jaaron, A. A. (2017). Assessing green human resources management practices in Palestinian manufacturing context: An empirical study. *Journal of Cleaner Production, 143*, 474–489.

McCarthy, J. (1989). Artificial intelligence, logic and formalizing common sense. In R. Thomason (Ed.), *Philosophical logic and artificial intelligence* (pp. 161–190). Dordrecht: Springer. Retrieved from www-formal.stanford.edu/jmc/ailogic.html.

Milliman, J., & Clair, J. (1996). Best environmental HRM practices in the U.S. In W. Wehrmeyer (Ed.), *Greening people: Human resources and environmental management* (pp. 49–74). Sheffield, UK: Greenleaf Publishing.

Mishra, P. (2017). Green human resource management. *International Journal of Organizational Analysis, 25*(5), 762–788.

Mukherjee, B., & Chandra, B. (2018). Conceptualizing green human resource management in predicting employees' green intention and behaviour: A conceptual framework. *Prabandhan: Indian Journal of Management, 11*(7), 36–48.

Naim, M. F., & Lenka, U. (2017). Linking knowledge sharing, competency development, and affective commitment: Evidence from Indian Gen Y employees. *Journal of Knowledge Management, 21*(4), 885–906.

Nilson, N. J. (1980). *Principles of artificial intelligence.* Palo Alto, CA: Tioga Publishing.

Nobari, A. R., Seyedjavadin, S. R., Roshandel Arbatani, T., & Rahnamay Roodposhti, F. (2018). Environmental concerns and green human resource management: A metasynthesis. *Iranian Journal of Plant Physiology, 8*(4), 2573–2576.

Obeidat, S. M., Al Bakri, A. A., & Elbanna, S. (2020). Leveraging "green" human resource practices to enable environmental and organizational performance: Evidence from the Qatari oil and gas industry. *Journal of Business Ethics, 164*(2), 371–388.

O'Donohue, W., & Torugsa, N. (2016). The moderating effect of 'Green' HRM on the association between proactive environmental management and financial performance in small firms. *The International Journal of Human Resource Management, 27*(2), 239–261.

Ojo, A. O., Tan, C. N. L., & Alias, M. (2020). Linking green HRM practices to environmental performance through pro-environment behaviour in the information technology sector. *Social Responsibility Journal.* https://doi.org/10.1108/SRJ-12-2019-0403

Pham, D. D. T., & Paillé, P. (2019). Green recruitment and selection: An insight into green patterns. *International Journal of Manpower, 41*(3), 258–272.

Pham, N. T., Hoang, H. T., & Phan, Q. P. T. (2019). Green human resource management: A comprehensive review and future research agenda. *International Journal of Manpower, 41*(7), 845–878.

Phillips, L. (2007). Go green to gain the edge over rivals. *People Management, 23*(9), 1–9.

Pinzone, M., Guerci, M., Lettieri, E., & Redman, T. (2016). Progressing in the change journey towards sustainability in healthcare: The role of 'Green' HRM. *Journal of Cleaner Production, 122*, 201–211.

Poba-Nzaou, P., Uwizeyemunugu, S., & Laberge, M. (2020). Taxonomy of business value underlying motivations for e-HRM adoption. *Business Process Management Journal*, *26*(6), 1661–1685.

Prakash, C., Saini, R., & Sharma, R. (2021). Role of Internet of Things (IoT) in sustaining disruptive businesses. In R. Sharma, R. Saini, C. Prakash, & V. Prashad (Eds.), *Role of Internet of Things (IoT) in sustaining disruptive businesses* (1st ed.). New York: Nova Science Publishers.

Rana, G., & Sharma, R. (2017). Organizational culture as a moderator of the human capital creation-effectiveness. *Global HRM Review*, *7*(5), 31–37.

Rana, G., & Sharma, R. (2019). Assessing impact of employer branding on job engagement: A study of banking sector. *Emerging Economy Studies*, *5*(1), 1–15.

Ren, S., Tang, G., & Jackson, S. E. (2018). Green human resource management research in emergence: A review and future directions. *Asia Pacific Journal of Management*, *35*(3), 769–803.

Renwick, D. W., Jabbour, C. J., Muller-Camen, M., Redman, T., & Wilkinson, A. (2016). Contemporary developments in Green (environmental) HRM scholarship. *International Journal of Human Resource Management*, *27*(2), 114–128.

Renwick, D. W., Redman, T., & Maguire, S. (2013). Green human resource management: A review and research agenda. *International Journal of Management Reviews*, *15*(1), 1–14.

Renwick, D. W., Redman, T., & Maguire, S. (2008). Green HRM: A review, process model, and research agenda. *University of Sheffield Management School Discussion Paper*, *1*, 1–46.

Roscoe, S., Subramanian, N., Jabbour, C. J., & Chong, T. (2019). Green human resource management and the enablers of green organisational culture: Enhancing a firm's environmental performance for sustainable development. *Business Strategy and the Environment*, *28*(5), 737–749.

Rubel, M. R. B., Kee, D. M. H., & Rimi, N. N. (2021). The influence of green HRM practices on green service behaviors: The mediating effect of green knowledge sharing. *Employee Relations: The International Journal*. https://doi.org/10.1108/ER-04-2020-+0163.

Sawang, S., & Kivits, R. A. (2014). Greener workplace: Understanding senior management's adoption decisions through the Theory of Planned Behaviour. *Australasian Journal of Environmental Management*, *21*(1), 22–36.

Sekhri, A., & Cheema, D. J. (2019). The new era of HRM: AI reinventing HRM functions. *International Journal of Scientific Research and Review*, *7*(3), 3073–3077.

Shah, M. (2019). Green human resource management: Development of a valid measurement scale. *Business Strategy and the Environment*, *28*(5), 771–785.

Sharma, R., & Rana, G. (2021). Revitalizing talent management practices through technology integration in industry 4.0. In R. Sharma, R. Saini, & C. Prakash (Eds.), *Role of Internet of Things (IoT) in sustaining disruptive businesses* (1st ed.). New York: Nova Science Publishers.

Sharma, R., Rana, G., & Agarwal, S., (2021). Techno innovative tools for employer branding in industry 4.0. In G. Rana, S. Agarwal, & R. Sharma (Ed.), *Employer branding for competitive advantage* (1st ed.). Boca Raton, FL: CRC Press. https://doi.org/10.4324/9781003127826-11.

Singh, A. (2014). Challenges and issues of generation Z. *IOSR Journal of Business and Management*, *16*(7), 59–63.

Siyambalapitiya, J., Zhang, X., & Liu, X. (2018). Green human resource management: A proposed model in the context of Sri Lanka's tourism industry. *Journal of Cleaner Production*, *201*, 542–555.

Song, W., Yu, H., & Xu, H. (2020). Effects of green human resource management and managerial environmental concern on green innovation. *European Journal of Innovation Management*.

Srivastava, A. P., Mani, V., Yadav, M., & Joshi, Y. (2020). Authentic leadership towards sustainability in higher education—An integrated green model. *International Journal of Manpower*, *41*(7), 901–923.

Stahl, A. (2021). How AI will Impact the Future of Work and Life. *Forbes*. Retrieved from https://www.forbes.com/sites/ashleystahl/2021/03/10/how-ai-will-impact-the-future-of-work-and-life/?sh=5f0b85ae79a3

Strohmeier, S., & Piazza, F. (2015). Artificial intelligence techniques in human resource management—A conceptual exploration. In *Intelligent techniques in engineering management* (pp. 149–172). Cham: Springer.

Tang, G., Chen, Y., Jiang, Y., Paille, P., & Jia, J. (2018). Green human resource management practices: Scale development and validity. *Asia Pacific Journal of Human Resources*, *56*(1), 31–55.

Tang, G., Yu, B., Cooke, F. L., & Chen, Y. (2017). High-performance work system and employee creativity. *Personnel Review*, *46*(7), 1318–1334.

Teixeira, A. A., Jabbour, C. J. C., & de Sousa Jabbour, A. B. L. (2012). Relationship between green management and environmental training in companies located in Brazil: A theoretical framework and case studies. *International Journal of Production Economics*, *140*(1), 318–329.

Tomer, G., & Rana, G. (2020). Green human resource management: A conceptual study. In Shivani Agarwal, Darrell Norman Burrell, & Vijender Kumar Solanki (Eds.), *Annals of Computer Science and Information Systems Volume 24, Proceedings of the International Conference on Research in Management & Technovation* (pp. 109–114), Adarsh Vidya Mandir's KDM Girls College, Nagpur, MH, India.

Voss, P. (2007). Essentials of general intelligence: The direct path to artificial general intelligence. In B. Goertzel & C. Pennachin (Eds.), *Artificial general intelligence: Cognitive technologies* (pp. 131–157). Berlin, Heidelberg: Springer.

Wee, Y. S., & Quazi, H. A. (2005). Development and validation of critical factors of environmental management. *Industrial management & data systems*, *105*, 96–114.

Wehrmeyer, W. (Ed.). (1996). *Greening people: Human resources and environmental management*. Sheffield, UK: Greenleaf Publishing.

Wong, M., Gardiner, E., Lang, W., & Coulon, L. (2008). Generational differences in personality and motivation: Do they exist and what are the implications for the workplace? *Journal of Managerial Psychology*, *23*(8), 878–890.

Yong, J. Y., & Mohd-Yusoff, Y. (2016). Studying the influence of strategic human resource competencies on the adoption of green human resource management practices. *Industrial and Commercial Training*, *48*(8), 416–422.

Yong, J. Y., Yusliza, M. Y., & Fawehinmi, O. O. (2019a). Green human resource management: A systematic literature review from 2007 to 2019. *Benchmarking: An International Journal*, *27*(7), 2005–2027.

Yong, J. Y., Yusliza, M. Y., Jabbour, C. J. C., & Ahmad, N. H. (2019b). Exploratory cases on the interplay between green human resource management and advanced green manufacturing in light of the Ability-Motivation-Opportunity theory. *Journal of Management Development*, *39*(1), 31–49.

Yusliza, M. Y., Othman, N. Z., & Jabbour, C. J. C. (2017). Deciphering the implementation of green human resource management in an emerging economy. *Journal of Management Development*, *36*(10), 1230–1246.

Zaid, A. A., Jaaron, A. A., & Bon, A. T. (2018). The impact of green human resource management and green supply chain management practices on sustainable performance: An empirical study. *Journal of Cleaner Production*, *204*, 965–979.

Zibarras, L. D., & Coan, P. (2015). HRM practices used to promote pro-environmental behavior: A UK survey. *The International Journal of Human Resource Management*, *26*(16), 2121–2142.

Zoogah, D. B. (2011). The dynamics of Green HRM behaviors: A cognitive social information processing approach. *German Journal of Human Resource Management*, *25*(2), 117–139.

2 The Role of Artificial Intelligence in Blockchain Applications

Haru Hong Khanh and Alex Khang PH

CONTENTS

2.1 Introduction 20
2.2 Artificial Intelligence 20
    2.2.1 AI-Based Systems in Theory Classification 20
        2.2.1.1 Reactive AI Machines 20
        2.2.1.2 Limited Memory AI Machines 21
        2.2.1.3 Theory of Mind AI Machines 21
        2.2.1.4 Self-Aware AI Machines 21
    2.2.2 Types of Artificial Intelligence 21
        2.2.2.1 Artificial Narrow Intelligence 21
        2.2.2.2 Artificial General Intelligence 22
        2.2.2.3 Artificial Super Intelligence 22
        2.2.2.4 Hybrid Artificial Intelligence 22
    2.2.3 AI-Based Systems in Industries 23
2.3 Blockchain Technology Overview 24
    2.3.1 The Generations of Blockchain 24
    2.3.2 Blockchain Architecture and Layers 25
        2.3.2.1 Blockchain Architecture 25
        2.3.2.2 Blockchain Layers 29
    2.3.3 Types of Blockchain 31
        2.3.3.1 Public Blockchain 31
        2.3.3.2 Private Blockchain 32
        2.3.3.3 Consortium Blockchain 33
        2.3.3.4 Blockchain Summary 33
2.4 The Role of AI in Blockchain Applications 34
    2.4.1 Data Process Flow Diagram 34
    2.4.2 Role of AI in Blockchain Application 34
        2.4.2.1 AI in Smart Contracts 34
        2.4.2.2 AI in Datasets 34
        2.4.2.3 AI in Data Protection 34
        2.4.2.4 AI in Data Encryption Mechanisms 36
        2.4.2.5 AI in Blockchain-Based Identity 36

DOI: 10.1201/9781003145011-2

2.4.2.6	AI in Authentication and Authorization Functionality	36
2.4.2.7	AI in Search Engines	36
2.4.2.8	AI in Decision-Making	36
2.4.2.9	AI in Data Monetization	37

2.5 Future of AI and Blockchain 37
2.6 Conclusion 37
References 38

2.1 INTRODUCTION

Artificial Intelligence (AI) is not new; it is a branch of the computer science field of more than seven decades that encompasses other related fields such as machine learning (ML) and deep learning (DL). Since the term AI became popular in human life in the mid-1950s (Anyoha, 2017), humans have started to live in the age of many concepts related to technologies such as Big Data, data transformation, Internet of Things (IoT), cryptocurrency, blockchain, etc.—an age when we have the capacity to develop machines and tools to help people to handle daily activities.

Today, AI's potential and application for an immensity of sectors in human life is increasing constantly, even though it is still in a hybrid pattern as an emerging or middle-stage technology. Through techniques such as machine learning and neural networks, global tech companies are investing in developing machines to think and behave more like humans.

AI application has already successful in many industries such as manufacturing, technology, healthcare, logistics, banking, marketing, entertainment and education. We can continue to use the functions of AI-based systems to replicate or simulate human intelligence that can perform tasks requiring human intelligence such as planning, learning, reasoning, problem-solving and decision-making.

Currently, numerous AI-based blockchains of solutions and applications have been developed by researchers and corporations. However, the wide adoption of these solutions is an issue in resource-constrained AI and IoT ecosystems.

In this chapter, integration of the available AI-based solutions into blockchain solutions is discussed, and the limitations and challenges of AI implementation in different blockchain sectors are presented.

2.2 ARTIFICIAL INTELLIGENCE

2.2.1 AI-BASED SYSTEMS IN THEORY CLASSIFICATION

AI-based systems are classified into four types of machines: reactive AI machines, limited memory AI machines, theory of mind AI machines and self-aware AI machines.

2.2.1.1 Reactive AI Machines

Reactive AI machines are the oldest form of AI systems, and have extremely limited ability to form memories or to use past experiences to inform current decisions. A reactive AI machine does not have memory-based functionality.

2.2.1.2 Limited Memory AI Machines

Limited memory AI machines are machines that can look into historical data to make decisions. Today, most chatbots and virtual assistants for self-driving vehicles have been developed and driven by limited memory AI.

2.2.1.3 Theory of Mind AI Machines

Theory of mind AI machines are machines that—like people and other living things in the real world—can have thoughts and emotions that directly affect their own behavior. The key features of these AI-based systems are the ability to understand these components by interacting needs, thoughts, emotions, beliefs and feelings.

2.2.1.4 Self-Aware AI Machines

The next generation of AI development is a self-aware AI machine with human-like systems that will not only be able to evoke feelings and understand the behaviors as they interact, but also have desires and their own needs, beliefs and expectations.

2.2.2 Types of Artificial Intelligence

Today, the most popular classification in the AI science industry is the classification of technology into artificial narrow intelligence (ANI), artificial general intelligence (AGI) and artificial super intelligence (ASI), as shown in Figure 2.1.

Based on the theoretical background and science research, Figure 2.1 presents the types of artificial intelligence.

However, in some special industries, a new AI model is a combination of AGI, ANI, and ASI into an AI-based system called hybrid artificial intelligence (HAI).

2.2.2.1 Artificial Narrow Intelligence

ANI, the so-called weak AI or narrow AI, is the only type of AI-based systems that humans have applied successfully today. It is oriented and designed to perform singular autonomous tasks using human-like capabilities.

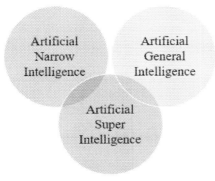

FIGURE 2.1 Three types of AI.

Basically, ANI can do nothing more than what they are programmed to do, and thus have a very limited or narrow range of competencies such as the functionality of searching on the internet, facial recognition, speech recognition, voice assistants, advertisement recognition or self-driving car, as in the following list.

Virtual Assistants
- Google search by Google
- Siri by Apple
- Alexa by Amazon
- Cortana by Microsoft

Software and Robots
- Image recognition software
- Facial recognition software
- Disease mapping software
- Prediction software
- Manufacturing robots
- Drones
- Industrial robots
- Self-driving cars

In other words, ANI has been impacting workforce performance and our expected outcome of using ANI is a *human-like machine*.

2.2.2.2 Artificial General Intelligence

AGI is referred to as "strong AI" or "deep AI"; it is an AI-based system of a machine that can think, understand, distinguish, mimic, behave and act in a way that can solve any problem like a human being in any situation.

AGI is based on the theory of mind AI machine as mentioned previously, which refers to the ability to discern feelings, emotions, beliefs, needs, thoughts and desires of intelligent entities. Today, it is easy to find AGI frameworks in training machines to understand human thinking and speculation.

AGI has been impacting human behavior, and our expected outcome of using AGI is an individual human-like brain.

2.2.2.3 Artificial Super Intelligence

ASI has probably marked the pinnacle of AI development, which has become by far the most intelligent on earth, and AI technology is not only already started to apply in many industries, but has also reshaped many models of services and productions up to now and will be changing business in the next decade.

ASI has been impacting human behavior, and our expected outcome of using ASI is a social human-like brain.

2.2.2.4 Hybrid Artificial Intelligence

HAI is designed and programmed to complete multiple tasks with one or many kinds of virtual assistant systems from Siri assistants to self-driving cars and autonomous weapons.

The Role of AI in Blockchain Applications

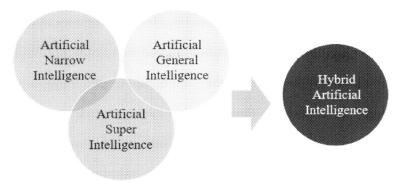

FIGURE 2.2 Hybrid AI model.

In the future, nobody can be sure if ASI or HAI will not be the last inventions of humanity. In fact, the fields of AI have become much broader than just the pursuit of human-like intelligence. Therefore, some AI scientists currently plan to develop a new version of the AI or combine AI types into one model, as shown in Figure 2.2.

Figure 2.2 demonstrates a hybrid artificial intelligence model that combines three popular AI models into an AI solution for the real world.

2.2.3 AI-Based Systems in Industries

In most cases in industries, AI primarily comprises business transactions and AI algorithms over IoT devices to each node, or cloud, from where the data can be accessed remotely. Most industrial applications popular for AI are as follows.

- Manufacturing industry
- Smart city industry
- Smart agriculture industry
- Healthcare industry
- Weapons industry
- Telecommunications industry
- Education industry
- Oil and gas industry
- Power industry
- Retail industry
- Supply chain industry

The mobility and density of IoT, eIoT (energy IoT), iIoT (industrial IoT) and aIoT (AI of Things) devices in large systems such as smart city systems may lead to an increase in network complexity and a decrease in its performance.

However, despite the complexity of implementation and operation, AI-based technologies have gained tremendous popularity in operations of heavy industries, smart cities, telecommunications and military networks.

2.3 BLOCKCHAIN TECHNOLOGY OVERVIEW

Blockchain technology was launched in 2008 and was created by the hidden persons behind the online cash currency Bitcoin, under the pseudonym of Satoshi Nakamoto (Wikimedia Foundation, Inc., 2010), a well-known Japanese name.

Today, as predicted by a scientist, blockchain is not only a digital currency, but it also has the potential to grow to be a data hub of the global record-keeping systems for corporations and governments.

The core concept of blockchain is a decentralized and distributed ledger technology (DLT) that is used for recording transactions and monitoring digital assets. The official definition of a blockchain is the two important components of a distributed ledger with smart contracts.

- A ledger is used to record transactions that capture all changes to a set of business objects. In a distributed ledger system, organizations collaborate with each other to maintain a consistent copy of a replicated ledger in a process that is called consensus or agreed-correct transaction.
- A smart contract is defined by the rules and conditions to query a ledger and generate new transactions that are recorded in it. Smart contracts describe the lifecycle of one or many business objects that are stored on the ledger with the time and author of creation, updates, and query activities.

In system architecture, a blockchain is a special type of database. In order to be able to understand blockchain, it is necessary to first understand what a database actually is.

2.3.1 THE GENERATIONS OF BLOCKCHAIN

If you are researching, reading, and following news of banking, financing, investing and cryptocurrency during recent years, you would have heard the term of the blockchain, a technology for using the record-keeping technology behind the digital currency networks.

Blockchain technology is still an important development, but we can divide the history of blockchain into at least four important generations, as shown in Table 2.1.

TABLE 2.1
The Generations of Blockchain

Generation	Year	Description	Use Cases
1st generation	1991–2008	Store and transfer of value based on coins with proof of work	Bitcoin, Ripple
2nd generation	2008–2014	Blockchain 2.0 based on coins with tokens	Ethereum
3rd generation	2014–2015	Cryptocurrencies are based on proof of stake, off-chain route, graph chain or centralized authorities	Enterprise blockchain
4th generation	2015 onwards	High scalability	Relational blockchain

2.3.2 BLOCKCHAIN ARCHITECTURE AND LAYERS

2.3.2.1 Blockchain Architecture

The blockchain architecture consists of five elements: *nodes*, *blocks*, *transactions*, *mining* and *consensus*.

2.3.2.1.1 Nodes

In simple terms, every computer or smart device in a blockchain network is a node and there exist several types of blockchain nodes.

As a definition, blockchain is a decentralized network and there is no central authority, so there is great value for blockchain nodes; each of them requires specific software and hardware configurations to get hosted or connected to others as shown in the blockchain nodes diagram in Figure 2.3.

Figure 2.3 presents a decentralized blockchain network based on blockchain nodes with different software and hardware configurations.

Fundamentally, there are two types of nodes. The first type is called full nodes and the second type is called lightweight nodes; these types include a constellation of a variety of nodes that are grouped to a cluster.

- A full node is a node that can keep and maintain a copy of the blockchain.
- A lightweight node does not keep a copy of the blockchain.
- The relationship between a full node and a lightweight node is the lightweight node must connect to a full node before it can interact with the blockchain system to send a transaction from the source node to the target node.

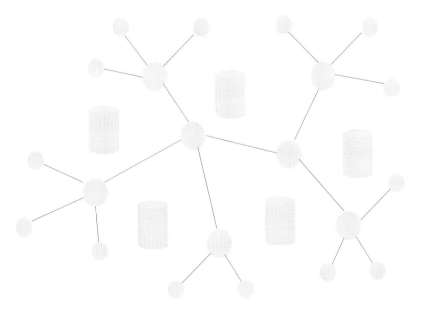

FIGURE 2.3 Blockchain nodes diagram.

2.3.2.1.2 Blocks

A block is a dataset and contains two main parts. The first part is valuable information as a block data header (block metadata will help to verify a transaction in a blockchain network), and the second part is the list of transactions as shown in the block structure diagram in Figure 2.4.

Based on the theoretical cryptocurrency and block structure, Figure 2.4 presents the diagram of key components in a blockchain transaction.

Block data header is defined in the block metadata which can be contained, such as the following.

- Time of the block creation.
- Version of the block creation.
- Previous block header hash.
- Encrypted hash of all transactions.
- nBits as a base-256 number.
- Nonce ("a number used once"), an arbitrary number that is assigned by a block creator (or a random number can be used just once in a cryptographic communication).

A rest data of a block will contain all the transactions that the miner has chosen, including a created block as the linked blocks diagram in Figure 2.5.

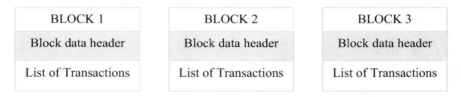

FIGURE 2.4 Block structure diagram.

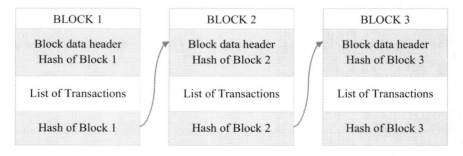

FIGURE 2.5 Linked blocks diagram.

The Role of AI in Blockchain Applications

Figure 2.5 presents the link between two blocks. The arrows display the direction of the relationship between block 1 and block 2 via hash data.

2.3.2.1.2.1 Transactions Blockchain transactions usually consist of a recipient address, a sender address and respective values, and they sound like a standard transaction that you would find on a statement such as ATM or credit card payments.

- As in Figure 2.5, a list of transactions in a block is established by the mining tools using by the miners, and blocks are divided into different types based on transaction functionalities.
- A blockchain transaction just changes the state of the consensus. This means the nodes independently hold their own copy of the block, and the current known "state" is calculated by processing each transaction in the order existing in the blockchain network.
- Blockchain transactions always contain one or many inputs and outputs with their value, and each input always references a previous transaction's output as an illustration of linked transactions in Figure 2.6.
  - An input value is value and a reference to an output value from a previous transaction.
  - An output value specifies a value and its address.

The difference between blockchain transactions and cryptocurrency transactions is that while blockchain transactions update the state of blocks by changing their relevant information, cryptocurrency transactions move some Bitcoin value from one address to another address.

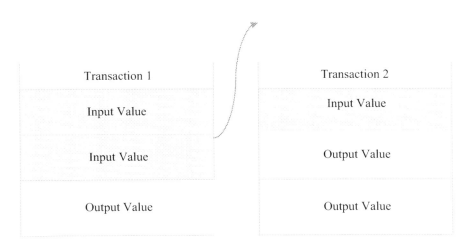

FIGURE 2.6 An illustration of linked transactions.

Figure 2.6 presents the link between two transactions in blockchain technology; the arrow displays the direction of the relationship between transaction 1 and transaction 2 via output and input data.

2.3.2.1.3 Mining

Mining is the process of working seamlessly in a blockchain network to create a valid block that will be accepted by the rest of the mining miners as a mining process shown in Figure 2.7.

In order to understand the mining process of blockchain technologies, Figure 2.7 illustrates the process of working seamlessly in a blockchain network.

The mining process involves hashing a block by checking to compare if the hash matches the current difficulty rules, and changing the nonce in the block header in case of mismatch, then hashing the block again.

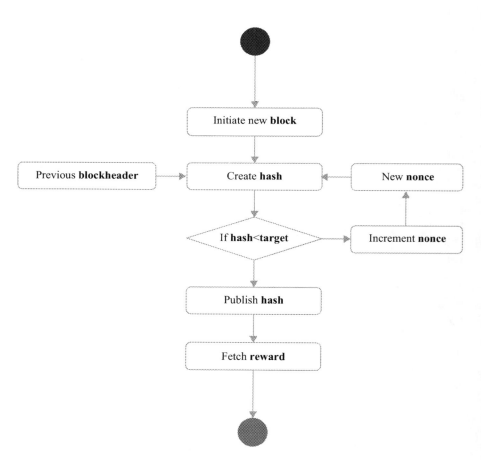

FIGURE 2.7 Mining process diagram.

The Role of AI in Blockchain Applications

- Blockchain block is a slight difference of nonce number (mentioned in the definition of blocks section) between owners.
- Miner is a node; they support the network by creating new blocks or forwarding information and maintaining a copy of the block in the blockchain network.

2.3.2.1.4 Consensus

- A consensus model is a fault-tolerant mechanism that is applied in blockchain systems to achieve the necessary agreement on a unique state of the network among multi-agent systems or distributed processes for record-keeping in other things.
- Consensus is the methodology that symbolizes the agreement of a maximum of nodes on the network having the same blocks in a validated blockchain package. In other words, it could be interpreted as a series of policies or conventions that each block will be harmony self-enforced in self-control mode.
- With a non-stop increase of the miners and nodes in the blockchain network, once the node has been created, it would link to an overall consensus updating with newer nodes. Currently, there are four types of consensus mechanism algorithms used for public and private blockchain applications, working on different principles in cryptocurrency and blockchain technologies.
  - PoW (proof of work) is a basic consensus algorithm used by the most popular cryptocurrency networks like Dogecoin, Litecoin and Bitcoin. PoW requires a node of participants to prove that the work completed and submitted by them qualifies them to receive the right to add new transactions to the cryptocurrency network.
  - PoS (proof of stake) is a popular consensus algorithm that evolved as an optimization alternative to the PoW algorithm. It involves the allocation of responsibility in maintaining the public ledger to a node of a participant in proportion to the number of cryptocurrency tokens held.
  - PoC (proof of capacity) allows sharing of memory or space for contributing to the nodes in the blockchain network. One node that has much memory or disk space would have granted more rights to maintain the public ledger.
  - BFT (Byzantine fault tolerance) is the property of a system that is able to resist the class of failures derived from the general problems in BFT systems, and this means that a BFT system is able to continue operating even if some of the nodes fail or act maliciously.

2.3.2.2 Blockchain Layers

Blockchain technologies are currently quite complicated in modeling and algorithmic attributes (McKinsey & Company, 2018), and blockchain is also not accepted as commercial payment in a few countries in the world, so it has been requiring years to develop better and apply to many primary industries.

30 Reinventing Processes Through AI

In order to develop a blockchain application, you can first simplify the discovery and better understand the seven layers of software application in the open systems interconnection (OSI) model, as Table 2.2 shows.

Second, dependent on the type of application, blockchain will have different application architecture as shown in the blockchain layer diagram in Table 2.3, so

TABLE 2.2
Demonstration of OSI Model

#	OSI Layer	TCP Layer	Data Transmission	Protocol	Hardware
7	Application layer	Application	Meaningful data (data from/to end user)	HTTP, FTP, IRC, SSH, DNS	
6	Presentation layer		Translation, encryption and compression of data	SSL, FTP, IMAP, SSH	
5	Session layer		Encrypted or decrypted data	NetBIOS, APIs, SOCKETs	
4	Transport layer	Transport	Segments	TCP, UDP, SCTP, DCP	
3	Network layer	Internet	Packages	IPv4, IPv6, IPSec (logical address)	Routers, switches
2	Data link layer	Network (coax, fiber, NICs, wireless)	Frames	MAC and LLC (physical address)	Ethernet, switches, bridge
1	Physical layer		Bit (transmit raw bit stream over physical network)		Repeater, modem, cables

TABLE 2.3
Demonstration of Blockchain Application Layers

Layer	Layer Name	Description	Activities and Methodologies
6	Application and presentation layers	**Raw data** Cryptocurrency, asset management, security settlements	User interface, smart contracts, hyperledger chain code
5	Execution layer	**Crypto and grant data** encrypt, compiler, docker	Encrypt, decrypt, compiler, docker
4	Data modeling layer	**Data structure** Block transactions	Hash, digital signature, Merkel tree
3	Consensus layer	**Blockchain transactions** Mechanism of transactions	PoW, PoS, PoC, BFT
2	Blockchain network layer	**Blockchain network layer** Blockchain transactions	Peer-to-peer (P2P) network
1	Hardware and infrastructure layer	**Foundation layer** Transport layer, network layer, data link layer, and physical layer	Virtual machine, docker, container, services, messaging

The Role of AI in Blockchain Applications

you need to discover and understand the architectural layers that you can build a blockchain application.

2.3.3 Types of Blockchain

Categorizing blockchain technology into two types of public blockchain or private blockchain is necessary for identifying the scope of use, type of users and the characteristics of blockchain applications. Currently, there are mainly the following three types of blockchain applications.

2.3.3.1 Public Blockchain

Public blockchain is known as permissionless or so-called non-permission blockchain. It is fully distributed, with no restrictions on participation, and encourages more participants to join the network. Therefore, nobody will have complete control over the blockchain network.

In the other words, a public blockchain is decentralized and does not have a single entity that controls the blockchain system. So, the data of a public blockchain are secure, as it is not possible to modify or update data once they have been validated on the blockchain network, as illustrated in the process diagram in Figure 2.8.

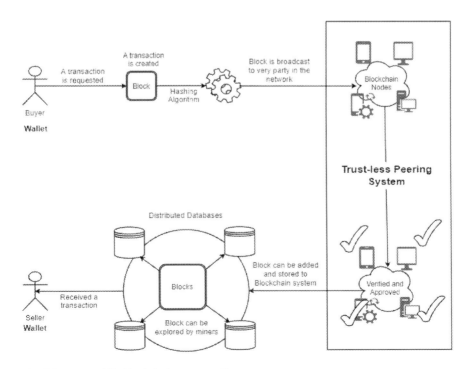

FIGURE 2.8 Public blockchain process diagram.

Figure 2.8 presents the life cycle of the transaction to ensure proper implementation and integrity of the smart contracts with trustless peering system from buyer to seller in the public blockchain ecosystem.

Nowadays, the most well-known and successful public blockchain applications are Bitcoin, Ethereum, Litecoin, Dogecoin, and PiCoin. In fact, public blockchain systems do not require any maintenance. That is why decentralized applications do not cost too much to create and operate.

2.3.3.2 Private Blockchain

Private blockchain is known as a permissioned blockchain. It is mainly operated and controlled internally by financial organizations, banks or governments, and also requires the participants to be invited and verified before they can be a member of the blockchain system.

In this context, all the transactions are shared and granted only to participants who are a part of the blockchain ecosystem; an illustration of the process is shown in Figure 2.9.

Based on the methodology of private blockchain technology obtained, Figure 2.9 presents the life cycle of the transaction to ensure proper implementation and

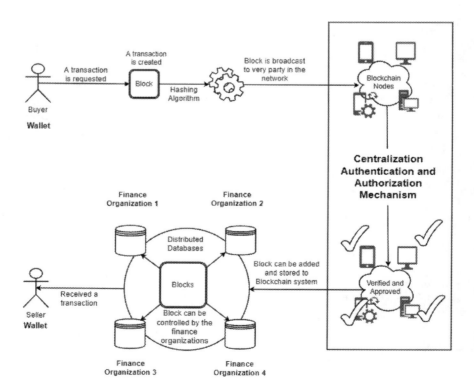

FIGURE 2.9 Private blockchain process diagram.

The Role of AI in Blockchain Applications 33

integrity of the smart contracts from buyer to seller in the private blockchain ecosystem.

One of the advantages of private blockchain is that it is more centralized than public blockchain technology; it can be governed and regulated by the association of financiers, banks or companies who can make sure that controllers are guiding participants.

In addition, a private blockchain can have a token—or may not have one, based on the policies or rules of the blockchain owner.

2.3.3.3 Consortium Blockchain

As with public blockchain, consortium blockchain is considered to be a fully decentralized system, with control over the blockchain not being in the hands of any single individual or entity. While private blockchain systems have the advantage of security, they are also slower and more costly because of the need for only certain verified individuals to perform any activity.

A consortium blockchain is a so-called federated blockchain or hybrid blockchain; it is considered as partially decentralized, and it combines the beneficial attributes of both public and private blockchain. Consortium blockchain has the advantages of a public ledger and it is operated under the leadership of a group instead of a single entity, which is a key characteristic of private blockchain.

2.3.3.4 Blockchain Summary

Experts say that private blockchain is the easiest platform to control and modify because it outweighs the public blockchain in that the organization can revert transactions, modify balances, change operation rules, etc. Therefore, the private blockchain model is suitable for any organization that is not ready to relinquish control such as financial companies, governments, etc.

In order to help you see the characteristics of each blockchain type, Table 2.4 shows all components, participants, permissions and relations in the blockchain system.

TABLE 2.4
Summary of Blockchain Types

	Public	Private	Consortium
Access	Anyone	Single organization	Multiple organizations
Participants	Anonymous	Known entities and with permissions	
Consensus Determination	Any node (or miner)	Single organization	Predefined group of nodes
Data Immutability	Rollback is required	Low, as rollback is possible	
Infrastructure	Decentralized	Centralized	Partially Decentralized
Security	Public	Private	Consortium
Transaction Speed	Slow	Faster	Faster
Efficiency	Low	High	High
Network Scalability	High	Low to medium	

2.4 THE ROLE OF AI IN BLOCKCHAIN APPLICATIONS

Nowadays, within the cryptocurrency context, popular and wide adoption in the age of AI engineering, building an efficient network communication and an excellent consensus mechanism are vital for an AI-based blockchain ecosystem.

Despite AI and blockchain technologies being the hottest technology trends at this time, and their highly different developing parties and applications, scientists and researchers have been looking for ways to combine them together.

2.4.1 DATA PROCESS FLOW DIAGRAM

In a blockchain network, each transaction issued by the smart contract must adopt the distributed ledger system; its encrypted and stored data inside each block depends on the type and structure of the blockchain; and it maintains data about the receiver, sender and the number of coins. Considering stages in blockchain application, you can apply AI technology as in Figure 2.10.

Based on the definition of the public and private blockchain models, Figure 2.10 presents the data process flow between buyer and seller wallets with the mediating role of the financial organizations.

As a data flow diagram, the role of AI in the blockchain is the engine that will enable involvement, analytics, management, protection and decision-making from encrypted data by hashing the distributed, decentralized, immutable ledger system.

2.4.2 ROLE OF AI IN BLOCKCHAIN APPLICATION

The convergence of blockchain and AI is obvious, as both technologies involve sensitive data and valuable assets. First, you will consider and explore what suitable stages can be applied to AI. Second, you can collaboratively implement stages in the data process flow of blockchains. Following are the stages to which AI technology can be applied.

2.4.2.1 AI in Smart Contracts

In smart contracts (Sandner et al., 2020), when two parties want to execute a secure, trusted transaction without the use of an intermediary, this is the ideal stage where AI technology could be used for blockchain application.

2.4.2.2 AI in Datasets

If blockchain contains Big Data and massive resources, AI technology focuses on fast processing and complex insights. By applying AI algorithms to the distribution of transactions or accessing to or out distributed ledgers, you can solve the current bottleneck for serving data more quickly.

2.4.2.3 AI in Data Protection

AI systems can help keep what is happening in the transaction lifecycle more secure, more reliable and more efficient. It can look for patterns and abnormal cases in

The Role of AI in Blockchain Applications

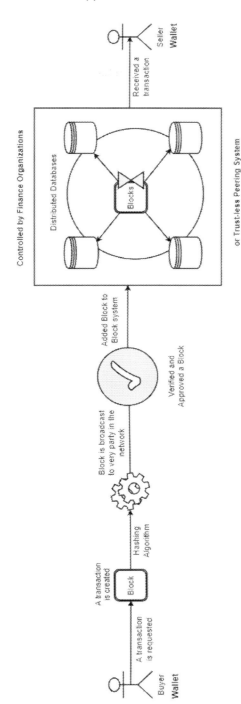

FIGURE 2.10 Data process flow diagram.

transactions being stored and distributed being performed on a particular node, and it can be used to alert parties when something may be happening.

2.4.2.4 AI in Data Encryption Mechanisms

As previously explained, once a transaction is created and distributed to blockchain system, the sensitive data need to be encrypted or decrypted, leaving them vulnerable to attack from hackers.

However, blockchain developers could solve these issues with the help of new methods of encryption that allow handling mathematical operations to the encrypted data and receive the encrypted result in whole lifecycle of transaction by the power of AI implementation.

Moreover, a branch of AI is computational intelligence, which can play many important roles in modern cryptographic systems. It can create more robust ciphers and improve the blockchain system's recovery via system attack-defense processes.

2.4.2.5 AI in Blockchain-Based Identity

In the stage of digital signature recognition for two parties before setting up the smart contract, the application of AI technology is a good example of object detection technology by using computer vision and image processing that can detect the instance of semantic objects of a class in digital images and videos.

2.4.2.6 AI in Authentication and Authorization Functionality

For blockchain consensus decisions and authentication and authorization functionality, or peering verifications as almost all data rely on the structured type, we are better off delegating decisions to AI functionality because AI technology is not prone to human behaviors and cognitive bias.

2.4.2.7 AI in Search Engines

Search functionality applies not only to social networks but also applies in blockchain applications or finance systems in the real world, where there are massive users. In a private or public blockchain system, it has always been a key technology and secret methodology for broadcasting the blocks to the parties in the blockchain network.

Since the proliferation of transactions and the increasingly complex use of encrypted datasets, search functionality has become more important, and blockchain owners are turning to artificial intelligence to improve the capabilities of their search services.

2.4.2.8 AI in Decision-Making

The global cryptocurrency market's fast variability involved in decision-making is increasing every day, and investors are always facing decision-making fatigue in transactions. Applying AI to support the decision-making process easier and faster includes understanding offering and accepting prices by sellers and buyers.

2.4.2.9 AI in Data Monetization

Blockchain applications can perfectly use AI technology to facilitate the sharing of data used across platforms. A great example is the use of machine learning models for making recommendations based on what you viewed in online retail or electronic news sites that you can see as smart advertisements on your mobile device or computer every day. TradingView (2021) has recommendations of cryptocurrencies in the real world.

2.5 FUTURE OF AI AND BLOCKCHAIN

In an era of industrial revolution 4.0, a consistent combination of several emerging technologies such as IoT, eIoT, aIoT, AI, cloud computing, cryptography algorithms, network models, and security mechanisms helps the blockchain systems can be encapsulated, authenticated, approved and distributed blocks across a network to each user as well as tied together by using complexity cryptographies that make the altering of these records almost impossible (Rossow, 2018).

In the industrial revolution 4, blockchain technology is not only using in the majority of finance and banking applications since it provides transparency, high-level trust mode, data security, and transactional accuracy and also applying in the industry of the smart city, e-commercial, military, healthcare, supply chain, energy management, logistics, politics, education, transportation, real estate, music, construction, and so on.

In the near future, humans will develop and interact with more features to prove blockchain technology is one of the best and lowest costs technologies using in across a wide range of other industries.

2.6 CONCLUSION

The future of convergences shown to explore some new solutions in AI and blockchain ecosystems will not only open new opportunities for effective performances, but will also permit stakeholders to protect ownership and control on their digital assets more security.

AI-based blockchain systems are believed to be able to assist a payment or business process such as a supply chain that involves many parties requiring trust and transparency, as well as efficiency in inter-party blockchain transactions, smart contracts, security of sensitive information and digital assets management.

Regarding the potential finance technologies, the emergence of AI technology as the next generation of blockchain platforms is predicted to transform business operations dramatically in every industry. Therefore, it is important to consider the methodology for embracing it, as there are many questions that need to be answered as this ecosystem progresses. Especially while most organizations and governments are planning in favor of the monopoly blockchain ecosystem, the challenges and barriers—either technical or even solutions—should not be underestimated from AI-based blockchain.

We would like to share the integration concepts in the meaningful embedding of these two technologies that could propel blockchain technology to a substantially higher level of the opportunities and challenges that can protect transactions from prime breaches and vulnerabilities in the era of cryptocurrency.

Notably, these pilot solutions have deployed to various blockchain platforms with different AI-based models, and given the competition between finance organizations and government cryptocurrencies, these models have not been reported in detail or published in the academic literature.

Eventually, the value propositions of AI-based systems were assessed, and the fields of this combination reconfirmed the application of blockchain technology to reduce infrastructure costs, boost the service quality and transaction governance, and increase competitiveness across the entire industry.

REFERENCES

Anyoha, Rockwell (2017). *The History of Artificial Intelligence*. Can Machines Think? Retrieved from https://sitn.hms.harvard.edu/flash/2017/history-artificial-intelligence/.

McKinsey & Company (2018, September). *Notes from the AI Frontier Modeling—The Impact of AI on the World Economy*. Retrieved from https://www.mckinsey.com/~/media/McKinsey/Featured%20Insights/Artificial%20Intelligence/Notes%20from%20the%20frontier%20Modeling%20the%20impact%20of%20AI%20on%20the%20world%20economy/MGI-Notes-from-the-AI-frontier-Modeling-the-impact-of-AI-on-the-world-economy-September-2018.ashx.

Rossow, Andrew (2018 April). *Bringing Blockchain into Industry 4.0*. Retrieved from www.forbes.com/sites/andrewrossow/2018/04/11/bringing-blockchain-into-industry-4-0/?sh=7ccf95e6dc7d.

Sandner, Philipp, Jonas Gross, and Robert Richter (2020, September). *Convergence of Blockchain, IoT, and AI*. Frankfurt School blockchain Center, Frankfurt School of Finance & Management, Frankfurt, Germany. Retrieved from www.frontiersin.org/articles/10.3389/fbloc.2020.522600/full#B22.

TradingView (2021, April). *Cryptocurrency Market*. Retrieved from www.tradingview.com/markets/cryptocurrencies/.

Wikimedia Foundation, Inc (2010, December). *Satoshi Nakamoto*. Retrieved from https://en.wikipedia.org/wiki/Satoshi_Nakamoto.

3 The Rise of Artificial Intelligence in Modern Healthcare Sector

Yogesh Pant and Balaji Dhanasekaran

CONTENTS

3.1 Introduction 39
    3.1.1 Overview of AI 40
    3.1.2 AI in the Healthcare Sector 42
3.2 Brain–Computer Interfacing 44
    3.2.1 Requirement for BCI Systems for Physically Disabled Patients 46
    3.2.2 Different Types of Brain Waves 47
    3.2.3 BCI-Controlled Wheelchairs 47
        3.2.3.1 Software Requirements 47
        3.2.3.2 Hardware Developments 48
        3.2.3.3 Working 49
3.3 Intelligent Vision 49
    3.3.1 Facial Recognition 51
    3.3.2 Object Recognition 51
    3.3.3 Currency Recognition 52
    3.3.4 Machine Learning Programming Using Python 53
        3.3.4.1 Libraries of Python 53
        3.3.4.2 Haar Cascade and Its Classifiers 54
        3.3.4.3 Comparison between Classifiers 55
        3.3.4.4 Facial Recognition Library 56
References 59

3.1 INTRODUCTION

Global technology organizations like Google, IBM, Babylon and many more are working in the healthcare sector using intelligent technologies. Researchers working in this domain have proposed many models which are interdisciplinary but beneficial in healthcare Musen et al. (2014). Google Inc. and collaborating institutions have researched and shown that an AI-based system that uses thousands of images for training has achieved physician-level sensitivity in diagnosing referable diabetic retinopathy (Gulshan et al., 2016).

DOI: 10.1201/9781003145011-3

Deep learning models build by MIT and Jameel Clinic in MIT's Computer Science and Artificial Intelligence Laboratory (CSAIL) predict cancer risk using mammograms. The model "Mirai" trained over 200,000 examples. Mirai has three modeling phases. First, risk modeling includes assessing risk and determines preventive measures. Second, it includes non-image risk factors like hormonal variables and age. Third, it uses an adversarial scheme that learns from mammograms. The MYCIN system developed for medical computer–based medical consultation uses a knowledge base and artificial intelligence (Shortliffe, 2012).

3.1.1 OVERVIEW OF AI

Artificial intelligence is a technique implemented in a machine or software that makes it intelligent enough that it can work as efficiently as human experts. Medical science is a field of precision, accuracy and knowledge. So we can say an intelligent system having knowledge of medical science will definitely change the ways of diagnosis and explore more possibilities. Nowadays, drug discoveries and robotic assistance in high-risk surgery are using artificial intelligence. There are number of other domains in the medical and health sector where intelligent technologies are being used to get early and efficient results.

The term "artificial intelligence" was coined by John McCarthy at the Dartmouth Summer Research Project on Artificial Intelligence in 1955 (McCarthy et al., 2006), but the process of making intelligent system and developing mathematical models was introduced way before that. McCulloch and Pitts introduced a perception and neural network in 1942 that mimics the working of biological neurons. Evolution of AI is based on time frame including 1960 studies focused on knowledge representation and natural language, genetic algorithms in 1980 and the 1990 neural network implementation on complex problems.

Artificial intelligence is a superset under which machine learning, deep learning and neural network techniques work. Machine learning is a subfield of artificial intelligence that is widely used and implemented for solving complex problems; at same time, it needs tremendous amounts of data to train the machine before test and use. Following are the type of machine learning models that are used for solving different types of problems.

1. Supervised learning is a type of machine learning and uses training data to train the machine and find correlation between input and output in training phase. This method is termed as supervised because it perform training and testing of the data to maintain correct correlation. Classification shown in Figure 3.1 and regression are the methods which fall under supervised learning.
2. Unsupervised learning is a type of machine learning method with its aim being to find patterns from unlabeled data. Anomaly detection and clustering technique shown in Figure 3.2, and dimension reduction, are used under unsupervised learning.
3. Deep learning is based on deep neural networks. A neural network having a number of hidden layers to train the system, and input and output layers for implementation, is called a deep neural network.

Rise of AI in Modern Healthcare Sector

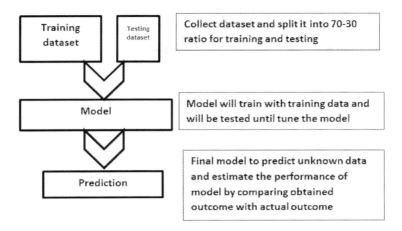

FIGURE 3.1 Supervised learning: regression.

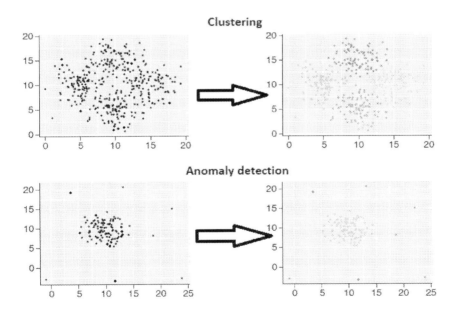

FIGURE 3.2 Unsupervised learning: clustering, anomaly detection.

A learning technique needs data to train the model and an interface to implement it. There are different types of learning mechanisms used in artificial intelligence based on available dataset, hypothesis and knowledge for problem-solving. Instance-based learning, inductive learning and analytical learning are those learning mechanisms. Based on problem domain and available data, we need to select an appropriate problem-solving technique to find optimized results. Regression, classification, decision trees and clustering techniques are some of the machine learning techniques.

3.1.2 AI IN THE HEALTHCARE SECTOR

According to Bini (2018), artificial intelligence is a tool that is used with cloud computing, the Internet of Things (IoT), mobile applications and Big Data. Combining technologies will result advanced and enhanced systems. Smartphone- and mobile-based patient monitoring systems are based on cloud computing and IoT that uses artificial intelligence (Rakshit & Sharma, 2020). Early disease detection based on magnetic resonance imaging (MRI) exams, blood samples or other techniques are based on deep learning models. Rule-based systems were successful in the 1970s with electrocardiograms (ECGs) and disease diagnosis.

Mobile-based healthcare applications are a trend that reduces crowds in the hospital and provides assistance and direct contact with doctor in the user's hand. Machine learning, cognitive intelligence and blockchain-based models provide the core functionality of AI in the application. It provides appointment booking, analyzing results and reports, direct contact with doctor via video call and payment method (Le Nguyen & Do, 2019).

Artificial intelligence is a mechanism to implement in a machine or software to make it as efficient as human being. Figure 3.3 has mentioned few of the implementation aspect of AI in healthcare.

Artificial intelligence is widely used in business processes, entertainment industry, agriculture, automobile industry, health care and many more. AI is used for data processing, optimization and searching Rana and Sharma (2019). Nowadays, AI-based inventions also make life easier for physically and visually disabled persons. Brain–computer interfacing (BCI) with AI is used for paralyzed patients to control home appliances and wheelchair. The same types of inventions which use image processing for facial recognition and object recognition based on machine learning model are used for visually disabled patients. We will explore both of these systems in this chapter.

Scientists and engineers have been working from more than the last five decades to improve speed and efficiency of computation. High speed encourages and expands the domain of implementation. The Australian company D-Wave System, large computer science and engineering companies like Google and IBM, and space research organization NASA all are working on next-generation computers called quantum computers because of their exponential potential of problem-solving speed. Medical science—especially drug discovery—will be totally changed in the quantum generation.

Liew (2018) proposed a roadmap related to implementation of AI in the field of radiology. Liew has worked for privacy, safety, moral and ethics. He has found that the overall cost of imaging decreases and throughput increases.

A system is being developed for controlling electrical home appliances on and off using brainwaves. NeuroSkyMind Wave software is used for brainwave sensing and transmission. To make it suitable for microcontroller use, raw brainwave data are fed into the MindWave MobileBCI device and put through a series of processing

Rise of AI in Modern Healthcare Sector

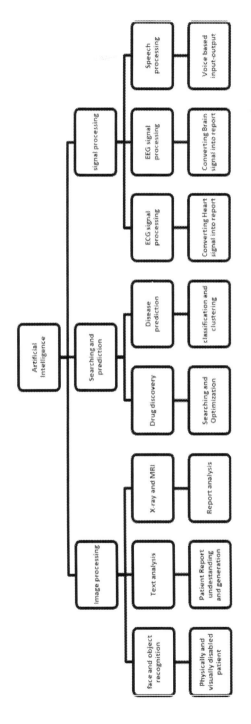

FIGURE 3.3 AI in healthcare.

algorithms. The resulting translated data is in cooperated into a microcontroller that has a digital device or electrical appliance connected to it. Using brainwaves, the user is able to control a device and turn the electrical appliance on and off.

Facial recognition technique is considered as a biometric technology to recognize people through real-time software or an application with the help of recorded datasets of images of people. OpenCV consists of face detection which is in-built. This face detector works on 90%–95% pictures of person who is in front of camera. At a particular angle, it is very difficult to recognize a person, so we require head pose estimation in 3D. But sometimes dim light of a picture, blurry picture and/or wearing glasses makes it a difficult task to detect a face. However, detection is quite more reliable than facial recognition, with accuracy of 80%–95%. Since the 1990s, the strong field which is use in research is facial recognition. By that time, there was no reliable and stable model created to recognize faces, but development of techniques occurs year by year. The beginning of facial recognition is done by extracting the coordinates of features of width of mouth, pupil and eye, and comparing them with measurements that are stored in datasets. Nowadays, there are many techniques, codes and algorithms which are used for facial recognition roaming around the whole world. Facial recognition is considered a very interesting topic of doing research. There are many published research papers which are related to this technology which include how facial recognition is done by extracting facial features, how facial recognition is to be implemented and algorithms for improvement. Facial recognition has become an important field of research because it has many applications like surveillance systems for criminals, ATM card owner identification, security monitors, taking attendance, and many more.

One of the enticements to making these investigations into personal identification has been to determine independent structures appropriate for hereditary investigation (Galton, 1888).

One more implementation aspect for facial recognition is intelligent vision for visually disabled patients. The robotics laboratory Willow Garage noticed in 2008 that there was a drastic growth in robotics capabilities; they become very advanced in commercial and research communities. From that time, Willow Garage with Vadim Pisarevsky and Gary Bradsky began to actively support OpenCV. Intel developed open source OpenCV library for computer vision through which programming for computer vision became so easy. OpenCV provides lots of functionalities for us like face tracking, face detection, facial recognition and various functions used in artificial intelligence. OpenCV is supported in Mac OS, Linux and Windows, making it a multi-platform framework. OpenCV has very understandable methods through which we can get satisfactory results.

3.2 BRAIN–COMPUTER INTERFACING

BCI is a technology that converts brain signals into digital form. After processing that digital data, we control things such as home appliances or wheelchairs for physically disabled persons. Analysis designing and implementation is a significant phase for understanding the concept of engineering. All the software or machines the world use are first designed and analyzed before implementation. A brain–computer interface (BCI) is a system that consists of hardware and software communications system. BCI is authorized to control external devices via cerebral activities.

Here, we review the state of the art of BCI systems, consisting of the following steps, which are also shown in Figure 3.4.

- Signal acquisition: collecting brain signals
- Signal enhancement: improving the signal strength
- Feature extraction: selecting appropriate signal patterns
- Classification: differentiating signals into classes based on patterns
- Control interface: triggering the external device based on class of signal generated

First, the signal acquisition step is implemented using neuro imaging modalities, each observing a dissimilar functional brain activity such as metabolic, magnetic and electrical activities. Second, sensing brain activity from different electrophysiological control signals that regulate user intentions. The third step is to deal with improving the performance used in the signal enhancement of the artifacts in the control signals. Fourth, the review studies some intelligent and mathematic algorithms used in the feature extraction and classification steps that are used to translate the information in the control signals into commands that instruct other devices. The goal of BCI systems is to deliver communications and access capabilities to severely disabled patients. For patients who are suffering from neurological neuromuscular disorders or even are totally paralyzed, BCI system provide them with access to and control of devices via brain activities. BCI systems are also helpful for brain stem stroke, amyotrophic lateral sclerosis or spinal cord injury patients.

The human brain comprises millions of neurons and an interconnected network of neurons. Brain–computer interface (BCI) uses mind wave sensors and neuron feedback. Thoughts and emotional states are represented by the interaction between these neurons. A unique electrical signal is generated based on muscle contraction. The state of human thoughts directly reflects different electrical waves. The control commands from brain are used to control external digital devices. To

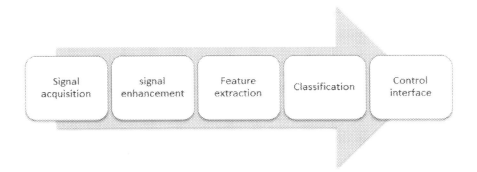

FIGURE 3.4 Brain–computer interface (BCI) system.

measure the living brain activity we need to use Electroencephalography (EEG) that provides measurement of electrical brain activity. Translation of different received brain patterns are used in BCI for uninterrupted communication between an external device and the living brain. To control the external device, EEG signals are needed. In this system, simple unipolar electrodes are used to record EEG signals from the patient's forehead using a headband with EEG electrodes and reference nodes. We have the signal attention and meditation level values. In addition to the raw dataset, we also extract eye-blinking signals. Therefore, attention and eye-blinking signals collected from EEG headset device are managed and sent through a Bluetooth interface at receiver end connected with microcontroller and then electrically interface with the wheelchair or home appliances to control them. Conversion of EEG activity cursor movement by a BCI also provides human–computer interfacing Fabiani et al. (2004).

3.2.1 Requirement for BCI Systems for Physically Disabled Patients

A general BCI framework proposed by Mason and Birch (2003) is a wonderful invention for physically challenged patients. It takes input from the human brain or eye movement, and converts that activity into a signal. Further, the received signals are used in microprocessors to perform activity using wheelchairs or to control home appliances. Figure 3.5 shows the modules used in this system from mounted headset to the appliance.

1. *Electroencephalography (EEG)* is a method used for measuring the brain wave activity as the brain's electrical signal. EEG is performed by placing electrodes on the patient's scalp. It is used to measure voltage variations within the neurons of the brain. EEG systems used in medical institutions for patient monitoring or research purpose are costly, but the development of cheaper and more consumer-friendly EEG devices for product development has put it in the mainstream market like EMOTIVE EPOC and brain sense.
2. *Electro-oculography (EOG)* is a system for measuring movements of the eye. Ocular region muscle-related movement can detect by EOG system by measuring the electrical changes such as eye movement and blinks.

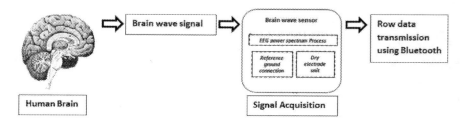

FIGURE 3.5 BCI data transmission.

Note: EEG = Electroencephalography

Rise of AI in Modern Healthcare Sector

3. *Bluetooth* is a wireless communication technology for short distances. In today's world, mobile cell phones are a necessity of everyone, and they have Bluetooth systems for data sharing. In BCI systems, Bluetooth is placed with EEG electrodes for sending the brain signals from the BCI device to a microcontroller placed on the appliance. Bluetooth devices have different classes; based on class, they have coverage limit from 10–100 meters.
4. *Microcontrollers* are programmed integrated circuits consisting of a memory unit, a processing unit and an input–output unit. They are small embedded computer systems which also provide interfacing with digital and analog sensors.

3.2.2 DIFFERENT TYPES OF BRAIN WAVES

In human brains, there are five types of brain waves. According to Picot et al. (2012), different brain waves have different frequencies, and the brain emits those in different states of mind described in Table 3.1.

3.2.3 BCI-CONTROLLED WHEELCHAIRS

3.2.3.1 Software Requirements

The ThinkGear connector application connects a BCI headset with the NeuroView application. NeuroView is a Windows application from neurosky, which is used to get the raw brain wave data, and the intelligent system of NeuroView shows the attention and meditation levels graphically. The software is based on the procedure to provide the control to operate patient's wheelchair that is as straightforward and simple within the boundaries of limited actions using NeuroView raw data (Bi et al., 2013).

Bharali et al. (2018) have proposed the following modes of operations:

- Standby
- Command
- Focus
- Running

These operations are sequential operation of the four different modes; each of the modes represent a state of the wheelchair. According to Jadhav and Momin (2018),

---

TABLE 3.1
Different Types of Brain Waves

Name of Brain Wave	State of Mind	Frequency
Delta wave	Deep sleep, loss of body awareness	<4 Hz
Theta wave	Deep meditation	4–8 Hz
Alpha wave	Physically and mentally relaxed; awake, but drowsy	9–13 Hz
Beta wave	Awake; normal alert consciousness	14–30 Hz
Gamma wave	Heightened perception	>30 Hz

first connect MindWave Mobile and Arduino's Bluetooth module placed on the wheelchair. Second, connect the Android MindWave mobile application with the headset, after establishing connection with the Android mobile application, and it starts fetching the signal quality value.

Signal quality is only shown in the mobile application while the device is placed by a user on their forehead and the reference electrode is placed in its right position. If the user is not wearing the headset, signal quality will be not detected. The headset electrodes should have proper contact with skin; if there is no contact or partial contact between the skin and the brain sense dry sensor, the application will show poor signal quality, and if the brain sense dry sensor has proper contact with the forehead, it will show good signal quality. To prevent unwanted action while signal quality is not good, an added safety precaution is there to send a stop command. When the signal quality becomes good, then the application again starts attending for any incoming blink data from the mobile application (Nielsen et al. 2006).

To implement wheelchair operation using the BCI control signal, the control signal is sent using Bluetooth attached to the headset and the signal has to be received in the mobile application; the control signal is then passed to another Bluetooth module attached to the wheelchair.

When a blink strength value received in the wheelchair module is above the threshold value, the MindWave application executes in cycling mode through values of forward, reverse, left and right. These direction signals each have a time span of ten seconds known as command mode, and a two-second intermission between each command mode for changing the direction value. To stop the cycling of direction, we have to provide two consecutive blinks, and whatever direction is shown in the cycle at the moment of the double blink event will become the chosen direction. To force stop, the cycling time elapsed between two blink events must be less than 400 milliseconds to be recognized by the brain sense device.

3.2.3.2 Hardware Developments

BCI-controlled wheelchair development needs a BCI headset, an Android mobile device and a wheelchair with Arduino microcontroller, Bluetooth and motor driver shown in Figure 3.6. In first module, the BCI headset consists of EEG electrode and a transmitting Bluetooth device; in second module, an Android application and Bluetooth device; and in third module, we have a wheelchair equipped with a microcontroller, motor driver and Bluetooth. The mobile application connects with headset Bluetooth module and wheelchair Bluetooth module simultaneously. First, the headset worn around the user's head will receive the raw brain wave data and wirelessly transmitting it through its own built-in Bluetooth module to the Android application. The Android application acts as a conduit between the MindWaveMobile application and the Arduino microcontroller placed in the wheelchair.

Brain wave data from the MindWaveMobile is not directly transmitted to the wheelchair microcontroller; the Android application filters the desired pattern and ensures that only relevant information is sent to the wheelchair microcontroller. The filtration or feature selection process has to be performed in the mobile application, and it reduces overhead from headset and microcontroller.

Rise of AI in Modern Healthcare Sector 49

The BCI headset has an adjustable headband. A power supply, a microcontroller, a dry electrode and a reference electrode are mounted on the headband. The position of the dry electrode is on the frontal head part on left side and reference electrode has to plug into the left ear.

The hardware structure for pilot testing a miniature wheelchair is having a HC-05 Bluetooth module to receive the control signal from Android application, a motor driver module that controls the movement of the wheel based on the instructions for direction and a microcontroller module that processes the signal received from Bluetooth and provide instruction to motor driver. The wheelchair is equipped with four motors and respective wheels. The motor driver has four direct current (DC) motors connected to it, and it is responsible for both its speed and direction control. The Android application plays the role of a conduit between the headset and wheelchair.

3.2.3.3 Working

- First, turn on the BCI headset module and connect it with the Android mobile application and wheelchair with Bluetooth.
- Trigger the command mode using force blinking.
- Once the direction cycles, blinking consecutively at least twice selects the direction and triggers focus mode.
- To get the robot running, try to focus on something to get focus level up to at least 50.
- Once the attention level reaches the threshold value of 50, running mode will be triggered and the robot will start running based on whatever direction has been selected earlier.
- Whenever needing to stop running mode, blink consecutively at least two times.
- If wanting to run it again, force blink once to trigger the operation cycle once again.

Controlling the robot using BCI headset command mode and direction cycle is proposed by Jadhav and Momin (2018).

3.3 INTELLIGENT VISION

Intelligent vision is a micro-controlled system that takes input using a camera and provides output as an audio signal to a visually disabled person. The image captured by the camera is processed by machine learning for image processing to identify surrounding objects. In this system, we have facial recognition, object recognition and currency recognition using machine learning and audio an output module, as shown in Figure 3.6.

This intelligent vision device is for visually challenged people, so this device contains multiple features. Our startup code does not contain anything, but is just to import files; whenever a user feels to use any of the features, they can directly call through it. We will only ask the user about the features they want in the startup code,

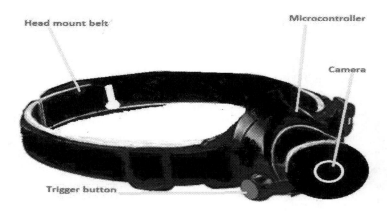

FIGURE 3.6 Head mounted intelligent vision.

and the startup code will do everything after that on its own. This device consists of a headband and camera module, microcontroller and control button mounted on it. The recognition processes for face, objects and currency are performed by a machine learning module. Different techniques of machine learning are used to implement these features.

Modules which our startup code contains are the following.

Module for Facial Recognition

This module contains the code of facial recognition. In this module, we do not have to pass any argument; after calling it, start using the camera for the picture and then it will predict according to the dataset, and after that it will return the name of the person (if known). You can also add an unknown person on runtime, but then you have to give the name of that person as a parameter so that device can recognize the person the next time it detects that person.

Module for Object Recognition

This module will also work the same as the facial recognition module; it also does not required any parameters and it will also work the same way as the facial recognition module, returning the name of object that has been captured after calling of the module. This module will use pre-trained libraries called "MobileNets"; using this, we can directly import the functions and can use them in our code. When we use this library and it returns some particular code which is not understandable by us directly, we have to use a separate file with names of objects called "COCO"; in this file name of objects exist which will returned by the code at the end as a result.

Module for Currency Recognition

This module will also work like other modules. It will return the value of the Indian denotation; this module also does not require any parameter. We

Rise of AI in Modern Healthcare Sector 51

require AI model for this feature which we are going to program or create through Google's "Teachable Machine." A teachable machine is something which helps to make AI models by simply providing images to the platform.

3.3.1 FACIAL RECOGNITION

Nowadays, there are many facial recognition algorithms and techniques available from developers around the world. According to Tsalakanidou et al. (2008), facial recognition becomes an interesting research topic and also has wide verity of applications in healthcare and industry. Different machine learning and neural network techniques are used to implement facial recognition.

Facial recognition is a significant research problem across different disciplines and numerous fields. Facial recognition initiates with feature extraction from the stored image. These features are width of eyes, pupil, mouth and other facial nodal points. The facial recognition system performs a comparison of a given face with patterns stored in the database, and returns the closest record (facial metrics). Facial recognition has numerous practical applications such as security, access control and surveillance systems; that is why in the last decade, a number of algorithms have been published and show improvement in terms of implementation and identification. A formal method of classifying faces was first proposed by Francis Galton (1888). Galton proposed a method for collecting facial profiles as face curves and finding their norm, and based on the deviations from the norm, classifies faces with other profiles. This type of classification is multi-model classification; it provides results in independent measures of vectors and those vectors could be compared with others in the stored database. Continuous progress in the algorithm and advancing technology has explored real-world application and implementation for facial recognition systems. Facial recognition in early stage works on detection of face patterns in cluttered scenes; those clustered scenes then normalize the face images to calculate the geometrical changes (Chellappa et al., 1995).

The appearance of facial patterns identifies the faces using appropriate classification algorithms, and the model-based schemes and logistic feedback is used for post processes. The availability of large databases of facial images, active development of algorithms and evaluation methods are the factors for rapid development and achievements of facial recognition. Facial recognition–related applications have to provide a still image or a video clip, and the system has to find the person available in the image or in the video.

This intelligent vision uses Python as a programming language and uses Open CV and facial recognition libraries with it to recognize the faces. Basically, in this project, the code finds the encodings of the known people images from the database which we have to provide, and after that—based on that encoding—the code predicts the name of the people captured in the camera at runtime.

3.3.2 OBJECT RECOGNITION

There are multiple studies on detecting and then recognizing objects from camera images. According to Dorner et al. (2015), object recognition is basically a technique of classifying objects based on images captured or live video streams. The authors of "Object Recognition by Effective Methods and Means of Computer Vision" designed new solutions for objects sensed by cameras to perform image recognition with the intention to detect and recognize objects. The camera module is used first to sense and receives objects, then determine and recognize them after prior digital processing of data provided by the camera. Neural network–based techniques like convolutional neural network (CNN) is widely used to perform the feature extraction that is the process of computer vision learning using neural-based methods. This type of recognition system may be used in various applications where the recognition and tracking of objects and understanding of the real world is required. Nowadays, intelligent vision system is also adapted by automobiles to recognize objects in front of vehicle.

Khurana et al. (2016) in "A Survey on Object Recognition and Segmentation Techniques" used images and videos to analyze object recognition and segmentation techniques. Object recognition techniques are being used by numerous applications such as robot navigation, human–computer interface, medical diagnosis, security, industrial inspection, autonomous vehicles and drones. Object segmentation is implemented in expanded fields such as video recognition, image processing, human activity recognition, gesture identification, shadow detection, and many more. Many authors have performed systematic analyses of several existing object recognition techniques and segmentation methods with respect to precise and arranged representation of parameters. In the field of object recognition, most of the used techniques are based on mathematical and algorithmic models.

3.3.3 CURRENCY RECOGNITION

Currency recognition which we are using is defined as the recognition of paper currency; here we are using this feature to predict the value of Indian paper currency with the help of cameras on runtime. Murthy et al. (2016) in "Design and Implementation of Paper Currency Recognition with Counterfeit Detection" used a method that can classify and subsequently verify Indian paper currency using a trained model of previously stored currency images and fundamental image processing techniques. Indian paper currency notes have different size and appearance based on their value.

MANI (Mobile Aided Note Identifier) is a system based on image processing; this application has launched by the Reserve Bank of India (RBI) for Android mobile users. The MANI application provides aid for visually impaired people to recognize Indian paper currency notes using their mobile phone camera module. The application does not need an internet connection to operate. The MANI application is user friendly, and also supports both English and Hindi languages. It gives output in the form of audio notification, and to support users with hearing impairments, the app has provide response in the form of predefined vibrations for different values of currency (Table 3.2). MANI is free to download and use. It is one of the best implementation of AI and mobile phone technology in healthcare for visually challenged and hearing impaired patients.

In this app, the user gets output with vibrations of the mobile, but we will give audio output from our device telling the value of the paper currency as described in

Rise of AI in Modern Healthcare Sector

TABLE 3.2
Predefined Vibration Alerts for Different Currency Denominations in MANI Applications

Currency	No. of Vibrations	Currency	No. of Vibrations
₹5	One	₹100	Five
₹10	Two	₹200	Six
₹20	Three	₹500	Seven
₹50	Four	₹2000	Eight

Table 3.2. In case the application is not able to identify the paper currency denomination, it uses a long vibration and the user has to scan the currency note again.

3.3.4 Machine Learning Programming Using Python

Python is an interpreter, high-level and object-oriented programming language. Python data structures, combined with dynamic binding and dynamic typing, makes it very attractive for RAD (rapid application development). Python language is easy to learn and simple syntax emphasizes readability. These features of language reduce the cost of program maintenance. Python is also used as scripting as well as a back-end language to connect existing components together. Python encourages code reuse and program modularity by supporting modules and packages. The Python interpreter and its extensive standard library are freely distributed and available in source or binary form without charge for all major platforms. Python libraries have prebuilt functions for artificial intelligence, machine learning, numeric calculations, graph plotting, computer vision and many more.

3.3.4.1 Libraries of Python

1. **OpenCV:** This is BSD-licensed product and makes it easy for implementation and modification in the code. OpenCV (open source computer vision library) is a machine learning software library and an open source computer vision. OpenCV is a very popular library built to provide a common infrastructure for computer vision applications. In Industry 4.0, computer vision is accelerating the use of AI and machine perception in the commercial products. We are using this library for performing image processing, video capturing for providing features to our users and performing multiple AI operations.

2. **Pyttsx3:** Pyttsx3 is a Python library for text-to-speech conversion. Unlike other libraries, Pyttsx works in offline mode, and is also compatible with both Python versions 2 and 3.

3. **Speech recognition:** We are using this library to perform "speech-to-text" operation in this system. Speech recognition refers to automatic recognition of human speech. It is an important part of HCI (human–computer interaction). If you have ever interacted with Alexa or ordered Siri to complete a task, you

have already experienced the power of speech recognition. These applications are proficient in speech recognition. A speech recognition system has a module for speech understanding and another module for speech generation.

4. **Facial recognition:** The library model of Python has an accuracy of 99.38% on the data of the "Labeled Faces in the Wild" benchmark. Recognizing and manipulating faces from Python is the simplest task, and it also works efficiently with single image labeled input. This library is the world's most used and simplest library. Facial recognition library is built using the C++ toolkit dlib's state-of-the-art machine learning with deep learning. Facial recognition library also provides a simple command line tool for facial recognition that lets you do facial recognition on a folder of images using command line. Depth information plays a vital role to improve the face detection system (Burgin et al., 2011).

We use Haar cascade classifiers given by OpenCV and the facial recognition library of Python to get the most accurate results of facial recognition. In this, we compare all classifiers of Haar cascade in terms of their accuracy of detecting faces. Classifiers are for eyes, nose, front-face, side-face and smile.

3.3.4.2 Haar Cascade and Its Classifiers

Haar cascade is an object detection algorithm used in machine learning (Padilla et al., 2012). It also helps in identification of objects whether the object is a non-living thing, a human being or something else.

According to Singh et al. (2013), Haar cascade generally contains four steps in its algorithm:

- Selection of features in object
- Data gathering (making datasets)
- Training of the model
- Give output (identifying object on the basis of datasets)

In the first step, we select particular features of an object by which we can identify that particular object. For example, In human face that features are eyes, nose, smile, whole front-face or side-face.

In the second step, with the help of computer vision (mostly via a camera), we take many images of the feature that we focused on in the previous step, keeping in mind that more images equals more accuracy during identification.

In the third step, we train the system depending upon the number of images that we take as input in data gathering.

In the fourth step, we define output as perfect if the system identifies an object on which it has been trained.

For this work, we take output of different classifiers of Haar cascade to compare their accuracy of identifying a particular face. These classifiers use features of the human face: eyes, nose, smile, front-face and side-face. To check accuracy at a particular time, we take particular number of images for one test then change numbers of images for another test (Le, 2011; Yang and Wang, 2016).

Rise of AI in Modern Healthcare Sector

3.3.4.3 Comparison between Classifiers

Test 1: We take 200 images of a person's particular features of face for making a dataset, then train our system and compare accuracy of all classifiers described in Table 3.3.

Test 2: We take 500 images of a person's particular features of face for making a dataset, then train our system and compare accuracy of all classifiers described in Table 3.4.

TABLE 3.3

Comparison of Classifiers When Number of Training Images Is 200

Classifier	Data gathering	Training dataset	Final testing	Accuracy
Eyes				-47%
Front face				40%
Smile				-44%
Side-face				42%
Nose				-88%

56   Reinventing Processes Through AI

TABLE 3.4
Comparison of Classifiers When Number of Training Images Is 500

Classifier	Data gathering	Training dataset	Final testing	Accuracy
Eyes				30%
Front face				43%
Smile				-15%
Side-face				43%
Nose				-88%

Test 3: We take 1,000 images of a person's particular features of face for making a dataset, then train our system and compare accuracy of all classifiers described in Table 3.5.

3.3.4.4 Facial Recognition Library

Facial recognition library is a Python library which is mainly used to recognize faces of different people. It is programmed in such a way that it automatically detects and recognizes faces; we just have to create dataset for it and import this library. The accuracy

Rise of AI in Modern Healthcare Sector

TABLE 3.5
Comparison of Classifiers When Number of Training Images Is 1,000

Classifiers	Data gathering	Training dataset	Final testing	Accuracy
Eyes				42%
Front-face				48%
Smile				1%
Side-face				46%
Nose				27%

of facial recognition via this library is noted as approximately 99% in many studies. It analyzes a live video stream which is played in front of a camera then matches that face to stored images that are known as image datasets. For this facial recognition, 128-d output feature vector is used to quantify a face. So, training can be by triplets.

For facial recognition library, we train the network by:

- Taking single image as input
- Classification of that image as output

So, by applying OpenCV and deep learning, we do the following steps by which process of facial recognition is done:

- Detection of face
- Computation of embedding of 128-d faces, through which we can quantify the face
- Train machine for embedding
- Recognition of faces takes place

For facial recognition library, we first create datasets of images; in the dataset, there is only a single image of a particular face used, as in Figure 3.7. Then, with the help of that single image as a dataset, it recognizes face shown in Figure 3.8.

According to the graph from Figure 3.9, we find that for Haar cascade, we have to take lots of images as inputs (image dataset) to get more accuracy during facial recognition. In comparison of classifiers, we get that accuracy of front-face and

FIGURE 3.7 Dataset having a single image.

FIGURE 3.8 Facial recognition.

Rise of AI in Modern Healthcare Sector

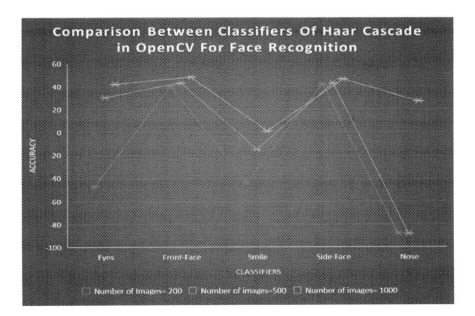

FIGURE 3.9 Comparison between classifiers of Haar cascade.

side-face classifiers are far better than any other classifiers regarding facial recognition. We also find that facial recognition library takes only a single input (image dataset) and gives more accurate results during facial recognition. We can also use it to identify faces in real time.

AI technology has proven its wide application domain in the healthcare sector for physically or visually disabled patients, and other application domains like drug discovery using pattern recognition and early stage disease detection using regression methods.

REFERENCES

Bharali, Sanjeeb, Mathi, Shyamala, Pooja, Dash, Chandra, Sharath, & Chavan, Bhushan. "A self-governing wheelchair for severely disabled patients in an indoor environment utilizing EEG and laser technologies." In *Emerging Trends and Innovations in Engineering and Technological Research (ICETIETR) 2018 International Conference*, pp. 1–5 (2018).

Bi, Luzheng, Fan, Xin-An & Liu, Yili. "EEG-based brain-controlled mobile robots: A survey." *IEEE Transactions on Human-Machine Systems*, 43(2), 161–176 (March 2013). doi:10.1109/TSMCC.2012.2219046.

Bini, S. A. "Artificial intelligence, machine learning, deep learning and cognitive computing: What do these term means and how will they impact health care?" *The Journal of Arthroplasty*, 33(8), 2358–2361 (2018).

Burgin, Walker, Pantofaruy, Caroline & Smart, William D. "Using depth information to improve face detection." In *2011 6th ACM/IEEE International Conference on Human-Robot Interaction (HRI)* (2011).

Chellappa, R., Wilson, C. L. & Sirohey, C. "Human and machine recognition of faces: A survey." *Proceedings of IEEE*, 83(5), pp. 705–740 (May 1995).

Dorner, J., Kozák, Š. & Dietze, F. "Object recognition by effective methods and means of computer vision." In *IEEE*, Strbske Pleso, Slovakia (June 9–12, 2015).

Fabiani, G. E., McFarland, D. J., Wolpaw, J. R. & Pfurtscheller, G. "Conversion of EEG activity into cursor movement by a Brain-Computer Interface (BCI)." *IEEE Transactions on Neural Systems and Rehabilitation Engineering*, 12(3), 331–338 (September 2004).

Galton, Francis. "Personal identification and description." *Nature*, 173–177 (June 21, 1888).

Gulshan, V. et al. "Development and validation of a deep learning algorithm for detection of diabetic retinopathy in retinal fundus photographs." *JAMA*, 316, 2402–2410 (2016).

Jadhav, N. K. & Momin, B. F. "Eye blink pattern controlled system using wearable EEG headband." In *2018 3rd IEEE International Conference on Recent Trends in Electronics, Information & Communication Technology (RTEICT), Bangalore, India*, pp. 2382–2386 (2018).

Khurana, Palak, Sharma, Anshika, Singh, Shailendra Narayan, & Singh, Pradeep Kumar. "A survey on object recognition and segmentation techniques." In *IEEE*, New Delhi, India (March 16–18, 2016).

Le, T. H. "Applying artificial neural networks for face recognition." *Advances in Artificial Neural Systems*, 2011, pp. 1–16 (2011).

Le Nguyen, T., & Do, T. T. H. "Artificial intelligence in healthcare: A new technology benefit for both patients and doctors." In *2019 Portland International Conference on Management of Engineering and Technology (PICMET)*, pp. 1–15 (2019).

Liew, C. "The future of radiology augmented with artificial intelligence: A strategy for success." *European Journal of Radiology*, 152–156 (2018).

Mason, S. G. & Birch, G. E. "A general frame work for brain-computer interface design." *IEEE Transactions on Neural Systems and Rehabilitation Engineering*, 11, 70–85 (March 2003).

McCarthy, J., Minsky, M. L., Rochester, N., & Shannon, C. E. "A proposal for the Dartmouth summer research project on artificial intelligence, August 31, 1955." *AI Magazine*, 27(4), p. 12 (December 2006).

Murthy, S., Kurumathur, J. & Reddy, B. R. "Design and implementation of paper currency recognition with counterfeit detection." In *2016 Online International Conference on Green Engineering and Technologies (IC-GET)*, Coimbatore, pp. 1–6 (2016).

Musen, M. A., Middleton, B. & Greenes, R. A. *Biomedical Informatics*. Shortliffe, E. H. & Cimino, J. J. (eds.), pp. 643–674 (Springer, London, 2014).

Nielsen, Kim Dremstrup, Cabrera, Alvaro Fuentes & do Nascimento, O.F. "EEG based brain computer interface—Towards a better control brain computer interface research at Aalborg university." *IEEE Transactions on Neural Systems and Rehabilitation Engineering*, 14(2), Article ID 1642769, pp. 202–204 (2006).

Padilla, R., Filho, C. C. & Costa, M. "Evaluation of Haar cascade classifiers designed for face detection." *Journal of WASET*, 6(4), pp. 323–326 (2012).

Picot, A., Charbonnier, S. & Caplier, A. "On-line detection of drowsiness using brain and visual information." In *IEEE Transactions on Systems, Man, and Cybernetics—Part A: Systems and Humans*, 42(3), pp. 764–775 (May 2012).

Rakshit, Pushpendu & Sharma, Ravindra. "Building Healthcare 4.0 with smart workforce." In *Proceedings of the International Conference on Research in Management & Technovation 2020*, pp. 91–94. PTI (2020).

Rana, G. & Sharma, R. "Emerging human resource management practices in Industry 4.0." *Strategic HR Review*, 18(4), 176–181 (2019).

Shortliffe, E. *Computer-Based Medical Consultations: MYCIN Vol. 2* (Elsevier, New York, 2012).

Singh, V., Shokeen, V. & Singh, B. "Face detection by Haar cascade classifier with simple and complex backgrounds images using opencv implementation." *International Journal of A dvanced Technology in Engineering and Science*, 1(12), pp. 33–38 (2013).

Tsalakanidou, F., Malassiotis, S., & Strintzis, M. G. "Face recognition." In Furht, B. (eds.), *Encyclopedia of Multimedia* (Springer, Boston, MA, 2008).

Yang, H. & Wang, X. A. "Cascade classifier for face detection." *Journal of Algorithms & Co mputational Technology*, 10(3), pp. 187–197 (2016).

4 Artificial Intelligence in Manufacturing

V. Harish, D. Krishnaveni, and A. Mansurali

CONTENTS

4.1 Introduction 63
    4.1.1 Literature Review 64
        4.1.1.1 Overview of AI—Technology 64
        4.1.1.2 Need for AI in the Manufacturing Domain 65
        4.1.1.3 AI in the Manufacturing Sector 65
        4.1.1.4 Advantages of Implementing AI in Manufacturing 66
        4.1.1.5 AI as a Catalyst to Smart Manufacturing 66
4.2 Objectives 66
    4.2.1 Advantages of Artificial Intelligence 67
    4.2.2 Disadvantages of Artificial Intelligence 68
        4.2.2.1 Shortcomings of AI 68
    4.2.3 Risks Associated with Artificial Intelligence 69
    4.2.4 AI in Process Capabilities 70
        4.2.4.1 Improvement at Process Level 70
        4.2.4.2 Benefits at Organizational Level 70
        4.2.4.3 AI—A Key Component of Manufacturing's Future 72
4.3 Conclusion 73
References 74

4.1 INTRODUCTION

Present generation manufacturing systems are becoming more complex and dynamic. Many manufacturing factories and businesses are facing increasing challenges due to a high number of uncertainties. Interdependencies between subsystems cause typical problems in a manufacturing system planning and control. The advent of artificial intelligence (AI) is defined as one of the most significant events in the recent times. Exponential development in the domain area of AI and machine learning shows great potential in transforming the manufacturing sector with the help of advanced analytics tools and algorithms to process the high amounts of manufacturing data collected. The pace of development in the field of AI particularly targeted toward the manufacturing sector is accelerating at a great speed every day.

We are on the verge of the tipping point of a new revolution, which is to be driven by rapid transformative and advancing technologies such as AI, three-dimensional (3D) printing, robotics, automation, and many more. These technologies are expected to

DOI: 10.1201/9781003145011-4

63

play a very crucial role in the future not only of manufacturing, but in other fields as well. Although the concept of artificial intelligence was discussed as early as in the 1940s by Alan Turing, its growth has been exponential in the last decade. Researchers who focused on AI such as Lin et al. (2017) mentioned that when AI is used properly and effectively, there is no need for fear about the adverse effects of AI. Smart and intelligent AI systems will improve the quality of life of humans in many aspects.

Artificial intelligence is the ability of a system to understand and analyze data, to learn and become familiar with it and to use the knowledge to perform assigned tasks and reach targeted goals with possibilities to adapt to situations presented. AI can help in reducing human error and also shift human focus to more productive areas. Smart manufacturing is noticeably projected among many business firms in recent years. The advancement is growing into a new developing area of "intelligent manufacturing" which clearly reflects the impact of AI smart technologies on the manufacturing sector. It is estimated that the "intelligent manufacturing" market will be over $200 billion in the year 2021 and is expected to increase to over $320 billion by 2022, showing annual growth rate of 12.5% (Dubey et al., 2020). Major companies like GE, Bosch and Microsoft started their investments in artificial intelligence long ago, the significant goal being use of machine learning–based programs to improve all aspects of manufacturing. The major reasons for using AI in manufacturing are to cut down labor costs, decrease product defects, reduce unplanned downtimes and also to increase productivity (Wang, 2019). While it might be possible for huge corporations to look into the possibility of using AI technology development strategy in manufacturing, this may not be so for smaller manufacturers as of now. Despite AI being a manufacturing sector revolution, it comes with its own disadvantages like high cost of implementation. AI is also generally thought of as lacking adaptivity and creativity. High growth in artificial intelligence is considered as a reason for global unemployment risk, but companies claim that it is a widespread misconception that AI will cause reduce human employment. Even though AI development is projected to beat a high pace, like many technologically disruptive innovations before, adoption of AI is slowed by concerns about costs, scalability and security.

This chapter aims to provide a bird's eye view of the future changes in the field of manufacturing due to the impact of the use of AI. The opportunities, threats and changes that one can expect are also touched upon.

4.1.1 Literature Review

4.1.1.1 Overview of AI—Technology

This section gives an overview about AI technology and the various aspects of AI in manufacturing sector.

The progress of AI is said to be one to the most significant events in recent past.

AI refers to "any computer system able to perform tasks that generally require human intelligence," according to the dictionary (Pillai et al., 2021). The ability of AI can be comprehended from three major sources:

- First, availability of a huge volume of data fueled by the "Internet of Things."

Artificial Intelligence in Manufacturing

- Second, the ability to perceive things; that is, the ability of the system to identify and recognize voice and images.
- Third, problem-solving ability; that is, the ability to continuously learn and improve.

AI is a pervasive technology, the use and impact of which expand across all sectors of the economy, ranging from manufacturing and healthcare to education to media. Adoption of AI can play an important role in forwarding sustainable development (Haenlein & Kaplan, 2019).

4.1.1.2 Need for AI in the Manufacturing Domain

AI is more relevant in the area of manufacturing because of the following reasons:

Shortage of expertise: One concern often cited by manufacturing companies is the shortage of required expertise (Chien et al., 2020). This is due to the fact that when workers retire or resign, they take away their experience and knowledge with them. The concern for organizations is not only to capture tacit experience and knowledge, but also to ensure that the knowledge is shared across different verticals within the organization (Li et al., 2017). Certain areas in which this problem could occur include production planning and process planning, detection or diagnosing issues in a machinery.

Highly complex situations: A major issue in manufacturing is the complexity involved in the decision-making process (Dubey et al., 2020; Rana et al., 2017). A firm will have lots of choices and opportunities to make decisions. The choices can be in designing of a product, production scheduling or inventory management. The constant changes in the needs of customers with very high expectations only adds to the many complex challenges involved.

High amount of information required: Due to the high awareness among customers, shortening lead time and varying requirements, a high amount of information needs to be shared among various supply chain partners (Kumar, 2017). The requirement to process this high level of information requires very high capabilities on the computing front (Cioffi et al., 2020).

4.1.1.3 AI in the Manufacturing Sector

AI in the manufacturing domain is expected to automate repetitive and boring human tasks, while factory workers will spend their time on more complex, creative and innovative tasks. Automation will not only substitute for labor, but also can complement labor and thereby increase productivity in many ways (Rana et al., 2019). AI collects virtual data as support for product development (Nti et al., 2021). AI virtual data can be used to monitor and improve shop floor performance. The manufacturing sector can use AI to identify unplanned downtime and accidents, and can also plan for machinery maintenance or repairs beforehand (Kusiak, 2017). Unstable price of raw materials is one of the challenges for manufacturers, and to combat this, AI algorithms can be used to predict material prices helping the firm to cope with unstable materials prices to remain competitive in the market (Lee et al., 2019). AI robots can be used as manufacturing robots for automating repetitive tasks and reducing human error. An AI robot can monitor its own accuracy and performance, and can train itself (Abd Aziz et al., 2021).

4.1.1.4 Advantages of Implementing AI in Manufacturing

Implementing AI in manufacturing will help organizations with defect detection as AI and self-learning capabilities can save many hours spent in quality control (Nikolic et al., 2017). Similarly, AI can be adopted for quality assurance such as in using image processing algorithms to spontaneously check whether an item has been perfectly manufactured (Kim et al., 2018). AI helps in assembly line integration by creating an interconnected application that extracts data from IoT connected equipment to ensure correct operation. Additionally, implementation of AI helps an organization generate designs for its products. The algorithm then checks all the possible outcomes of a solution and suggests or recommends design alternatives (Hayhoe et al., 2019).

Implementation of AI in Manufacturing Will Enable

- Mass customization (Wu et al., 2017).
- Implementation of robots (Sharp et al., 2018).
- Predictive maintenance that would help a firm to minimize its operation and maintenance costs (Nica et al., 2021).
- Software as a Service (SaaS): AI-based supply chain management solutions (Tao et al., 2018).
- Improve human–robot collaboration (Zhou et al., 2018).

AI and machine learning tools can create a unified ecosystem enabling continuous interchange of information. This makes the system self-sustaining, and also improves the process in every facet of supply chain management (Alonso, 2020).

4.1.1.5 AI as a Catalyst to Smart Manufacturing

AI can monitor thousands of data sources relating to manufacturing, with analysis of any news, event or incident that could affect production (Mao et al., 2019). It would help organizations to deepen their customer insight and understanding (Alexopoulos et al., 2020). Through AI, organizations can customize and create easy-to-understand dashboards for each customer. AI captures real-time alerts when any incident occurs during production (Ghobakhloo & Ching, 2019). It reduces the time taken to identify potential issues in production activity, enabling rapid response to address such issues before they escalate. AI quantifies and lowers the risk of incidents across the supply chain and during production (Ghahramani et al., 2020) It enables more informed decisions relating to the overall manufacturing process. Finally, AI helps to improve customer loyalty and the bottom line.

4.2 OBJECTIVES

- To understand the advantages of AI for manufacturing companies.
- To understand the disadvantages of adopting AI with respect to manufacturing companies.

Artificial Intelligence in Manufacturing

- To identify the risks associated with the implementation of AI in manufacturing companies.
- To develop a model to understand the capabilities of AI toward process development and other organizational benefits.

4.2.1 ADVANTAGES OF ARTIFICIAL INTELLIGENCE

Direct automation: The Internet of Things (IoT) connects all IoT-enabled gadgets organization-wide, be it the shop floor or the administrative office, leading to a pathway for effective automation linking various manufacturing processes (Acharya et al., 2021). AI-enabled "programmable logic controllers" linked with deep learning ability can make decisions automatically based on generated data and alter processes without any human intervention. These AI capabilities can greatly increase efficiency and effectiveness of a manufacturing company, and pave the way for managing processes remotely (Senthiil et al., 2019).

Uninterrupted production: Because of their needs and in certain cases due to legal requirements, humans are forced to take frequent breaks and can put up with only few hours of work a day (Yadav & Jayswal, 2018). In order to improve efficiency, a manufacturing firm has to be run around the clock, which is a challenge as companies might not have required number of resources or expertise. By going in for advanced robotics coupled with decision-making capability, AI can work on the assembly line 24 hours a day, seven days a week (24/7) (Dai et al., 2020). This makes it possible to increase production capabilities to a very great extent.

Improve efficiency: Adoption of AI allows a system to process massive volumes of data, which can be used to gain useful insights into manufacturing processes, machinery running, customer behavior, etc. (Dubey et al., 2020). Trends of the production process, better forecasts and clear business dynamics can all be estimated by adopting an efficient AI system over time. AI has the ability to forecast details, optimize processes, and monitor and control any kind of anomalies across the supply chain from supplier to end customer (Cioffi et al., 2020).

Improve safety: Humans are frail and are prone to making errors resulting in injuries on the factory floor and in the organizational environment, resulting in huge misery and financial claims. This is an issue that AI—in conjunction with robotics—can eliminate to a great extent (Nti et al., 2021). Working this way necessitates a reduction in human resources, especially when the nature of the work is hazardous or requires extraordinary effort. Also, AI can replace normal routine working practices, thereby reducing the number of industrial accidents and significantly increasing the safety of the organization and its resources (Pillai et al., 2021). Furthermore, the amalgamation of specific sensors with IoT devices will make the workplace safer and more efficient (Nti et al., 2021).

Less operational cost: Many businesses are cautious about incorporating artificial intelligence (AI) into the manufacturing process, as it usually necessitates a major financial investment (Morris et al., 2017), but studies indicate that the return on investment (ROI) is substantial and investing in AI is a great decision in the long run (Massaro et al., 2019). Businesses will benefit from dramatically decreased running costs and predictive maintenance.

Quality control: AI systems can be implemented across organizations to perform preventive maintenance on equipment and machinery (Aggour et al., 2019). AI systems will assist in anticipating malfunctions of equipment and failures of machines using machine learning, monitoring the performance of machines and understanding varying operating conditions, and can also take steps thereby preventing defects even before they happen. Effective AI systems can lead to faster and efficient inspections, which in turn can reduce unplanned downtime (Patel et al., 2021).

4.2.2 Disadvantages of Artificial Intelligence

AI maybe a new technology, but it cannot outperform humans because it cannot bring in inventions and can only do the work that it is programmed to do (Syam & Sharma, 2018). AI utilizes past information to read a general model or a pattern, which can be used to predict similar future occurrences. AI cannot think beyond a certain limit like humans can, and it has been identified that AI works best with human collaboration (Lee et al., 2019).

4.2.2.1 Shortcomings of AI

Cannot create: Even though AI can write papers or articles, reply to or compose emails, and even write full movie scripts, it can only adopt a formula and continue with the formula, while learning on the go. As of now, it lacks the imagination of a human and cannot formulate new ideas. AI cannot do something different beyond a pattern. An AI system cannot create an output of what it has not already seen or learned (Nasiri et al., 2017).

Cannot care: AI systems are target-oriented and do not consider the emotions of people, as unlike humans, they are generally focused on one target (Sun & Medaglia, 2019). Even though AI in many fields has a significantly reasonable rate of accuracy, it cannot feel the emotions of the humans it is chatting with by the way they are texting (Jha et al., 2019).

Irreplaceable Skills

Six skills will not be replaced by artificial intelligence.

- Ability to judge
- Empathy
- Creativity
- Ability to plan considering varying factors
- Physical skills

Security concerns and employee misgivings: Adoption of AI is often connected with the high potential risk it poses with respect to security measures. In some cases, even engineers who have developed the AI systems do not endorse the products for lack of complete understanding of the AI systems (Eren et al., 2021). Scientists who have been working on the AI front for decades have often been quoted that they have understood only the tip of what the AI system is capable of, and hence, it should not be meddled with.

Artificial Intelligence in Manufacturing   **69**

Implementation cost is high: The complexity of constructing AI-based systems often results in exorbitant costs. This is also due to the fact that AI technology is still in the emerging stages of development. As the machine becomes smarter by the day, this AI-based computer software needs regular updates to keep up with the demands of the changing world (Ashima et al., 2021). In the event that the program fails entirely, the procedure of retrieving missing data and the codes, and reinstalling the device, can be a nightmare for organizations.

Cannot replace humans entirely: Machines, coupled with AI, definitely and without any question are capable of working much more effectively when compared to humans. Even though it might be tempting to replace all the work done by humans with AI, existing technology is not adequate. No matter how intelligent or smart an AI system is, it will never be able to replace a human. We may be frightened at the prospect of being replaced by robots, but it is still a long way off (Baryannis et al., 2019). Machines are logical, but they lack feelings and moral principles. They lack the ability to form relationships with other people, which is a required skill for leading a group of people, although they can store high volumes of data, extracting the necessary information from them is a time-consuming process that pales in comparison to human intelligence (Bajic et al., 2018).

Unemployment risk is high: With fast and dynamic changes in the field of AI, one wonders when AI systems will be in a position to replace humans (Tao et al., 2018). AI is likely to perform a majority of routine and monotonous tasks that are often repetitive and do not involve decision-making. According to a McKinsey Global Institute report, smart AI systems and robots will replace close to 30% of the world's existing human labor by the year 2030. It stated that "between 400 and 800 million workers will be displaced by automation by 2030, causing as many as 375 million people to switch work categories entirely (McKinsey, 2017, p. 11)." As a result, it is possible that AI would result in less human interference, causing significant disruption in job standards. Nowadays, most companies are introducing automation at some level in order to substitute low-skilled employees with computers that can perform the same tasks more effectively (Makridakis, 2017).

4.2.3   Risks Associated with Artificial Intelligence

Automation-spurred job loss: AI is witnessing immense growth in the present world. The advancement of AI systems in every sector has affected the amount of manpower utilized in an organization. According to a two-year study by the McKinsey Global institution, there could be huge number of jobs lost by the year 2030 due to AI systems which will replace the repetitive and monotonous work performed by human labor (McKinsey, 2017).

Privacy violations: AI can solve many of the business challenges often faced by organizations, such as quickly spotting a few questionable charges among the thousands of invoices and the ability to connect the dots from the high volume of data available (Tao et al., 2018). Companies gather large amounts of data about customers and vendors on a daily basis. Using this data, AI can build profile data, often without the permission of the concerned person (Walczak, 2019).

Deepfakes: Deepfakes are advanced AI technologies which make it even more difficult to differentiate between original and fabricated media, resulting in a threat to people and organizations (Maras & Alexandrou, 2019). Deepfakes create or alter existing images, audio, and/or video, resulting in duplicates that look genuine. The rapid increase in the adoption of digital enablers such as video conferencing and other similar digital tools has increased the ability to perform malicious things (Westerlund, 2019).

Weapon automatization: If and when AI is permitted to handle nuclear and biological weapons, it will potentially be vulnerable to manipulation by enemies (Albahar & Almalki, 2019). The risk and potential of loss are significantly high.

4.2.4 AI IN PROCESS CAPABILITIES

AI offers huge potential to manufacturing firms, and based on a literature review, the following model (Figure 4.1) is proposed to understand the capabilities of AI. The capabilities of AI could have an impact on manufacturing sector from the process level dimension and organizational level.

4.2.4.1 Improvement at Process Level

The improvement at the process level at many organizations is measured using various indicators such as effectiveness, efficiency, value, etc. The data for these relevant metrics are collected, measured and analyzed.

At the process level, improvements can happen at the following three levels.

- Automation: Use of AI-based systems to replace human work and automate the existing process, thereby reducing or replacing human work. This could result in higher productivity and higher efficiency as machines can work consistently for a longer period of hours than can humans. Automation could result in increased efficiency and reliability, and provide routinization for manufacturing firms.
- Information: Using AI to collect, consolidate, store, retrieve and share information along the value chain. With the high and fast requirement of information on one side and the huge volume of data being generated on the other side, AI systems will enable organizations to improve their responsiveness, improve their quality of decisions and optimize in the area of resource management.
- Transformation: Using AI systems to have a transformation in the business process including re-engineering and remodeling a business setup. AI systems can play a critical role in the areas of re-engineering and design engineering, and can significantly improve the quality of services provided by the organization.

4.2.4.2 Benefits at Organizational Level

Financial: The common metrics used to see the benefits of AI from the financial perspective are cash flow, profitability, market value, etc. Organizations adopting AI

Artificial Intelligence in Manufacturing

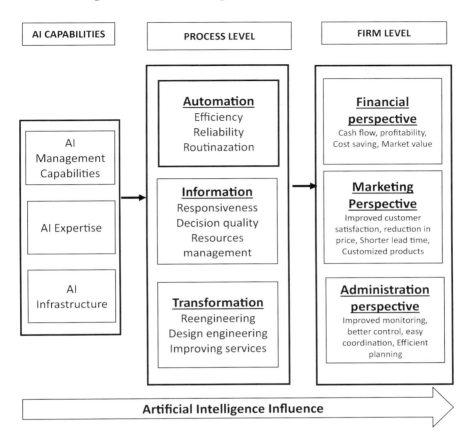

FIGURE 4.1 Proposed model to understand AI capabilities. AI capabilities refer to the organizational ability to adopt and use AI in value creation.

AI management capabilities refers to management's ability to use technology to create value for the organization. The potential is "strategic planning," resulting in better coordination and control.

AI expertise refers to the skillset and knowledge of the human resources of a firm specific to AI-related technologies.

AI infrastructure refers to the availability of all the resources—including the hardware, software, data, the required systems and network—for the effective implementation of an AI system.

can see benefits such as better cash flow due to better information flow and faster production, increased profitability due to enhanced productivity and increased market value.

Marketing: The benefits of AI from the marketing perspective can be seen in better customer satisfaction, reduction of price, shorter lead time and ability to offer customized products. Manufacturing companies adopting AI can access high information and use it to improve customer satisfaction. Shorter lead times and increased efficiency can result in reduction of price of products (Sharma et al., 2021). AI, coupled with robotics, could help companies to offer customized products to their customers.

For traditional manufacturers, integrating AI across everyday processes and equipment is not a small task. Corporations which continue with the status quo risk becoming obsolete. Considering the low level of adoption of technology, lack of skilled resources in the manufacturing industry has resulted in the sector struggling to keep up with the fast pace of digital transformation.

AI has been identified as one of the most disruptive technologies of present times. It will have an impact on end-to-end supply chains, but at the same time, it can be observed that manufacturing is one of the slowest sectors to adopt the digital way. Even though most decision-makers know it is important for their firm to implement AI, most of them are still in the pilot stage.

4.2.4.3 AI—A Key Component of Manufacturing's Future

AI helps in forecasting the malfunctioning of machines. With this, we will be able to predict beforehand and can repair the machines in order to maintain good efficiency. It also helps in predicting the requirements of customers and to help sales experts in managing their schedules. Any defects can be easily identified throughout the entire process. It can diminish the maintenance downtime. Areas where AI can play a vital role in manufacturing include the following.

- Predictive maintenance
  - Improved overall equipment efficiency (OEE)
  - Increased productivity
  - Improved safe working conditions

- Safe operational conditions
  - Replace humans in dangerous working conditions
  - Warn in advance of any likely mishap

- Improve productivity
  - Can work 24/7
  - Can collaborate or assist humans in performing tasks
  - Can replace with monotonous and repetitive tasks
  - Real-time production planning
  - Process optimization

- Improving quality
  - Continuous inspection
  - Increased consistency
  - Improved supply chain practices
  - Improved quality assurance

- Cost optimization
  - Optimal use of resources
  - Improved productivity

Artificial Intelligence in Manufacturing 73

- Improved quality
- Real-time supply chain
- Logistics optimization
- Improved inventory management

- Core manufacturing
  - Product development
  - Design customization
  - Factory performance improvement
  - Better forecast

- Faster information processing
  - Connected factories
  - Connected supply chain
  - Ability to process high volume of data and take faster decisions

Artificial Intelligence can assist in manufacturing in the following ways.

- AI can help manufacturing in real time by identifying production bottlenecks, tracking scrap rates in real time, meeting customer delivery dates, and more.
- AI can be implemented in manufacturing for maintaining machinery through preventive maintenance, thereby increasing overall equipment effectiveness.
- AI techniques can assist in working with various constraints and recommend optimized results.
- AI can be used to alert an operator if there are deviations in the production process.
- AI can help manufacturers comply with regulatory requirements.
- AI can help manufacturers improve demand forecast accuracy.
- AI can be used to calculate component images in the production line, thereby spotting any deviations in standards in real time.
- AI can predict—and thereby improve—worker safety, and achieve sustainability goals.
- AI can assist in designing new products in real time, thereby reducing the time to market for new products.
- AI can give a whole new meaning to "quality control" for manufacturers.

4.3 CONCLUSION

In today's fast-paced environment, it is important for manufacturers to respond quickly to changing consumer demands and maximize new market opportunities to sustain global competitiveness. In the current era, information technology, operation technology and global megatrends are on a collision course. This will drive changes in the way an organization conducts its operations and the way in which the organization interacts with its customers and suppliers. The manufacturing sector is considered a perfect fit for AI. Although Industry 4.0 is in its early stages, significant benefits from AI are being witnessed all over the world. From the design process and production floor to supply chain and administration, AI is changing the way organizations manufactures their products and materials. High revenue volatility, short production lead times, increased regulations and inspections, manufacturing capacity and supply chain demands drive the need to adopt AI in the manufacturing sector.

REFERENCES

Abd Aziz, N., Adnan, N. A. A., AbdWahab, D., & Azman, A. H. (2021). Component design optimisation based on artificial intelligence in support of additive manufacturing repair and restoration: Current status and future outlook for remanufacturing. *Journal of Cleaner Production*, 126401.

Acharya, S., Mishra, D., Rizvi, A., Haleem, A., Bahl, S., & Javaid, M. (2021). Artificial Intelligence (AI) and its applications in Indian manufacturing: A review. *Current Advances in Mechanical Engineering: Select Proceedings of ICRAMERD*, 2020, 825–835.

Aggour, K. S., Gupta, V. K., Ruscitto, D., Ajdelsztajn, L., Bian, X., Brosnan, K. H., . . . & Vinciquerra, J. (2019). Artificial intelligence/machine learning in manufacturing and inspection: A GE perspective. *MRS Bulletin*, 44(7), 545–558.

Albahar, M., & Almalki, J. (2019). Deepfakes: Threats and countermeasures systematic review. *Journal of Theoretical and Applied Information Technology*, 97(22), 3242–3250.

Alexopoulos, K., Nikolakis, N., & Chryssolouris, G. (2020). Digital twin-driven supervised machine learning for the development of artificial intelligence applications in manufacturing. *International Journal of Computer Integrated Manufacturing*, 33(5), 429–439.

Alonso, R. S. (2020, June). Deep tech and artificial intelligence for worker safety in robotic manufacturing environments. In *International Symposium on Distributed Computing and Artificial Intelligence* (pp. 234–240). Springer, Cham.

Ashima, R., Haleem, A., Bahl, S., Javaid, M., Mahla, S. K., & Singh, S. (2021). Automation and manufacturing of smart materials in additive manufacturing technologies using Internet of Things towards the adoption of industry 4.0. *Materials Today: Proceedings*, 45, 5081–5088.

Bajic, B., Cosic, I., Lazarevic, M., Sremcev, N., & Rikalovic, A. (2018). Machine learning techniques for smart manufacturing: Applications and challenges in industry 4.0. (p. 29). Department of Industrial Engineering and Management Novi Sad, Serbia.

Baryannis, G., Dani, S., & Antoniou, G. (2019). Predicting supply chain risks using machine learning: The trade-off between performance and interpretability. *Future Generation Computer Systems*, 101, 993–1004.

Chien, C. F., Dauzère-Pérès, S., Huh, W. T., Jang, Y. J., & Morrison, J. R. (2020). Artificial intelligence in manufacturing and logistics systems: Algorithms, applications, and case studies. *International Journal of Production Research*, 58(9), 2730–2731.

Cioffi, R., Travaglioni, M., Piscitelli, G., Petrillo, A., & De Felice, F. (2020). Artificial intelligence and machine learning applications in smart production: Progress, trends, and directions. *Sustainability*, 12(2), 492.

Dai, H. N., Wang, H., Xu, G., Wan, J., & Imran, M. (2020). Big data analytics for manufacturing internet of things: opportunities, challenges and enabling technologies. *Enterprise Information Systems*, 14(9–10), 1279–1303.

Dubey, R., Gunasekaran, A., Childe, S. J., Bryde, D. J., Giannakis, M., Foropon, C., . . . & Hazen, B. T. (2020). Big data analytics and artificial intelligence pathway to operational performance under the effects of entrepreneurial orientation and environmental

Artificial Intelligence in Manufacturing

dynamism: A study of manufacturing organisations. *International Journal of Production Economics*, 226, 107599.

Eren, B., Guvenc, M. A., & Mistikoglu, S. (2021). Artificial intelligence applications for friction stir welding: A review. *Metals and Materials International*, 27(2), 193–219.

Ghahramani, M., Qiao, Y., Zhou, M., Hagan, A. O., & Sweeney, J. (2020). AI-based modeling and data-driven evaluation for smart manufacturing processes. *IEEE/CAA Journal of AutomaticaSinica*, 7(4), 1026–1037.

Ghobakhloo, M., & Ching, N. T. (2019). Adoption of digital technologies of smart manufacturing in SMEs. *Journal of Industrial Information Integration*, 16, 100107.

Haenlein, M., & Kaplan, A. (2019). A brief history of artificial intelligence: On the past, present, and future of artificial intelligence. *California Management Review*, 61(4), 5–14.

Hayhoe, T., Podhorska, I., Siekelova, A., & Stehel, V. (2019). Sustainable manufacturing in Industry 4.0: cross-sector networks of multiple supply chains, cyber-physical production systems, and AI-driven decision-making. *Journal of Self-Governance and Management Economics*, 7(2), 31–36.

Jha, S., Sahai, T., Raman, V., Pinto, A., & Francis, M. (2019). Explaining AI decisions using efficient methods for learning sparse Boolean formulae. *Journal of Automated Reasoning*, 63(4), 1055–1075.

Kim, D. H., Kim, T. J., Wang, X., Kim, M., Quan, Y. J., Oh, J. W., . . . & Ahn, S. H. (2018). Smart machining process using machine learning: A review and perspective on machining industry. *International Journal of Precision Engineering and Manufacturing-Green Technology*, 5(4), 555–568.

Kumar, S. L. (2017). State of the art-intense review on artificial intelligence systems application in process planning and manufacturing. *Engineering Applications of Artificial Intelligence*, 65, 294–329.

Kusiak, A. (2017). Smart manufacturing must embrace big data. *Nature News*, 544(7648), 23.

Lee, W. J., Wu, H., Yun, H., Kim, H., Jun, M. B., & Sutherland, J. W. (2019). Predictive maintenance of machine tool systems using artificial intelligence techniques applied to machine condition data. *Procedia Cirp*, 80, 506–511.

Li, B. H., Hou, B. C., Yu, W. T., Lu, X. B., & Yang, C. W. (2017). Applications of artificial intelligence in intelligent manufacturing: A review. *Frontiers of Information Technology & Electronic Engineering*, 18(1), 86–96.

Lin, H. W., Tegmark, M., & Rolnick, D. (2017). Why does deep and cheap learning work so well? *Journal of Statistical Physics*, 168(6), 1223–1247.

Makridakis, S. (2017). The forthcoming Artificial Intelligence (AI) revolution: Its impact on society and firms. *Futures*, 90, 46–60.

Mao, S., Wang, B., Tang, Y., & Qian, F. (2019). Opportunities and challenges of artificial intelligence for green manufacturing in the process industry. *Engineering*, 5(6), 995–1002.

Maras, M. H., & Alexandrou, A. (2019). Determining authenticity of video evidence in the age of artificial intelligence and in the wake of deepfake videos. *The International Journal of Evidence & Proof*, 23(3), 255–262.

Massaro, A., Manfredonia, I., Galiano, A., & Xhahysa, B. (2019, June). Advanced process defect monitoring model and prediction improvement by artificial neural network in kitchen manufacturing industry: A case of study. In *2019 II Workshop on Metrology for Industry 4.0 and IoT (MetroInd4. 0&IoT)* (pp. 64–67). IEEE.

McKinsey Global Institute. (2017). Technology, jobs, and the future of work.

Morris, K. C., Schlenoff, C., & Srinivasan, V. (2017). Guest editorial a remarkable resurgence of artificial intelligence and its impact on automation and autonomy. *IEEE Transactions on Automation Science and Engineering*, 14(2), 407–409.

Nasiri, S., Khosravani, M. R., & Weinberg, K. (2017). Fracture mechanics and mechanical fault detection by artificial intelligence methods: A review. *Engineering Failure Analysis*, 81, 270–293.

Nica, E., Stan, C. I., Luțan, A. G., & Oașa, R. Ș. (2021). Internet of things-based real-time production logistics, sustainable industrial value creation, and artificial intelligence-driven big data analytics in cyber-physical smart manufacturing systems. *Economics, Management, and Financial Markets*, 16, 1.

Nikolic, B., Ignjatic, J., Suzic, N., Stevanov, B., & Rikalovic, A. (2017). Predictive manufacturing systems in Industry 4.0: Trends, benefits and challenges. *Annals of DAAAM & Proceedings*, 28.

Nti, I. K., Adekoya, A. F., Weyori, B. A., & Nyarko-Boateng, O. (2021). Applications of artificial intelligence in engineering and manufacturing: A systematic review. *Journal of Intelligent Manufacturing*, 1–21.

Patel, A. R., Ramaiya, K. K., Bhatia, C. V., Shah, H. N., & Bhavsar, S. N. (2021). Artificial intelligence: prospect in mechanical engineering field—A review. *Data Science and Intelligent Applications*, 267–282.

Pillai, R., Sivathanu, B., Mariani, M., Rana, N. P., Yang, B., & Dwivedi, Y. K. (2021). Adoption of AI-empowered industrial robots in auto component manufacturing companies. *Production Planning & Control*, 1–17.

Rana, G., Sharma, R., & Goel, A. K. (2019). Unraveling the power of talent analytics: Implications for enhancing business performance. In Rajagopal, Behl R. (ed.), *Business Governance and Society*. Palgrave Macmillan, Cham. https://doi.org/10.1007/978-3-319-94613-9_3.

Rana, G., Sharma, R., & Rana, S. (2017). The use of management control systems in the pharmaceutical industry. *International Journal of Engineering Technology, Management and Applied Sciences*, 5(6), 12–23.

Senthiil, P. V., Sirusshti, V. A., & Sathish, T. (2019). Artificial intelligence based green manufacturability quantification of a unit production process. *International Journal of Mechanical and Production Engineering Research and Development*, 9(2), 841–852.

Sharma, Ravindra, Saini, Ashwini Kumar, & Rana, Geeta. (2021). Big data analytics and businesses in Industry 4.0. *Design Engineering*, 2021(2), 238–252.

Sharp, M., Ak, R., & Hedberg Jr, T. (2018). A survey of the advancing use and development of machine learning in smart manufacturing. *Journal of Manufacturing Systems*, 48, 170–179.

Sun, T. Q., & Medaglia, R. (2019). Mapping the challenges of artificial intelligence in the public sector: Evidence from public healthcare. *Government Information Quarterly*, 36(2), 368–383.

Syam, N., & Sharma, A. (2018). Waiting for a sales renaissance in the fourth industrial revolution: Machine learning and artificial intelligence in sales research and practice. *Industrial Marketing Management*, 69, 135–146.

Tao, F., Qi, Q., Liu, A., & Kusiak, A. (2018). Data-driven smart manufacturing. *Journal of Manufacturing Systems*, 48, 157–169.

Walczak, S. (2019). Artificial neural networks. In *Advanced Methodologies and Technologies in Artificial Intelligence, Computer Simulation, and Human-Computer Interaction* (pp. 40–53). IGI Global.

Wang, L. (2019). From intelligence science to intelligent manufacturing. *Engineering*, 5(4), 615–618.

Westerlund, M. (2019). The emergence of deepfake technology: A review. *Technology Innovation Management Review*, 9(11).

Wu, D., Jennings, C., Terpenny, J., Gao, R. X., & Kumara, S. (2017). A comparative study on machine learning algorithms for smart manufacturing: Tool wear prediction using random forests. *Journal of Manufacturing Science and Engineering*, 139(7).

Yadav, A., & Jayswal, S. C. (2018). Modelling of flexible manufacturing system: A review. *International Journal of Production Research*, 56(7), 2464–2487.

Zhou, J., Li, P., Zhou, Y., Wang, B., Zang, J., & Meng, L. (2018). Toward new-generation intelligent manufacturing. *Engineering*, 4(1), 11–20.

5 Customer Behavior Prediction for E-Commerce Sites Using Machine Learning Techniques
An Investigation

G. Rajesh, S.P. Preethi, R. Shanmuga Priya, L. Rajesh, and X. Mercilin Raajini

CONTENTS

5.1 Introduction 80
5.2 Methodologies 80
    5.2.1 Collection of Data and Preprocessing 81
    5.2.2 Data Analysis 82
    5.2.3 Outlier Detection 82
    5.2.4 Training the Dataset 82
    5.2.5 Prediction 82
        5.2.5.1 Random Forest 83
        5.2.5.2 Support Vector Machine 83
        5.2.5.3 Decision Tree 83
        5.2.5.4 Logistic Regression 83
        5.2.5.5 Naive Bayes 83
        5.2.5.6 K-Means Clustering 84
5.3 Techniques Employed 84
    5.3.1 Machine Learning Approaches 84
    5.3.2 Data Mining Approaches 85
        5.3.2.1 Data Preparation 85
        5.3.2.2 Identification of Patterns 86
        5.3.2.3 Deployment 86
    5.3.3 Hybrid Model Construction Approach 86
    5.3.4 Neural Network Approach 87
    5.3.5 Individual- and Segment-Level Approaches 87

DOI: 10.1201/9781003145011-5

5.4 Performance Evaluation and Result..87
5.5 Conclusion 91
References..91

5.1 INTRODUCTION

The primary objective is to analyze the forecast of personalized customer behavior based on the individual level or segment level modeling approaches. The data collected in e-commerce sites are valuable resources to predict customer behavior, which acts as a vital task to adopt a technique to predict the sales trend. These analyses are helpful for organizations to determine what the customers want and what strategies can be adopted to make the customers stay with the organization. The customer behavior analysis at an individual level or as segment-based strategies are developed to satisfy the customer and make them come back for more.

Predicting the next basket balances the demand-driven supply chain of an organization. This forecast plays a vital role in some of the planning and decision-making of an enterprise. The market growth of an organization is highly dependent on predicting customer behavior accurately. Accurate prediction of customer behavior leads to effective revenue generation for the company. Analysis of customer behavior ensures the satisfaction of the customer and also increases the loyalty of the customers toward the enterprise. There have been more than 17,800 papers related to customer behavior analysis since 2016, of which the 21 most related papers were referred for this survey. This survey analyzes customer behavior data using various existing approaches such as machine learning, data mining and neural networks using a predictive model approach to recommend the next basket shopping list. It gives an evaluation method for validating and comparison of the result obtained by various techniques.

The predictive analysis methodologies are discussed in Section 5.2. Section 5.3 summarizes the techniques employed to forecast customer behavior. Evaluation of the performance and results derived from it is discussed in Section 5.4. Section 5.5 concludes that predictive modeling performance may differ with the employed machine learning algorithm and exhibits logistic regression as the best model in comparison.

5.2 METHODOLOGIES

The predictive analysis methodologies of data as shown in Figure 5.1 are collection and preprocessing, data analysis, outlier detection, training and prediction. The machine learning algorithms such as random forest, support vector machine, decision tree, logistic regression, naive Bayes and K-means clustering techniques, along with data mining, hybrid model, neural network and individual-and segment-level approaches, are employed to forecast customer behavior. Performance and results are computed along with the comparison were made between F-measure and mRHR value of individual-and segment-level approaches where logistic regression exhibits the best model results.

Customer Behavior Prediction

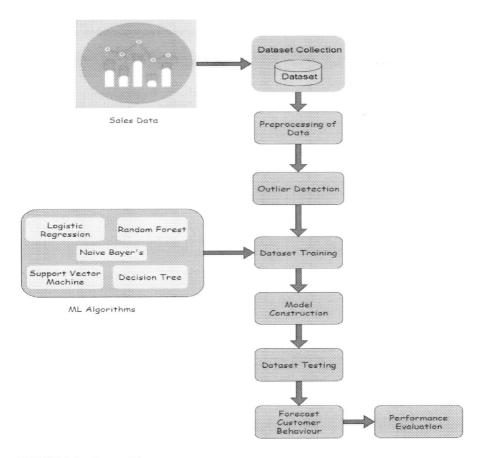

FIGURE 5.1 Scopus Diagram

5.2.1 COLLECTION OF DATA AND PREPROCESSING

The prediction dataset to be used is collected from the internet. Referred papers used datasets such as the grocery chain's loyalty card system dataset (Peker et al., 2017; Peker et al., 2018), datasets from Kaggle on customer behavior (Harsh Valecha et al., 2018), BigMart sales dataset (Kumari Punam et al., 2018; Rai et al., 2019), inventory dataset (Maheswari & Packia Amutha Priya, 2017; Li et al., 2020; Singh et al., 2020) and daily sales revenue dataset (Rankothge Gishan Hiranya Pemathilake et al., 2018; Bowen et al., 2020). Most influencing attributes are selected by using machine learning or data mining techniques from a large number of attributes in collected data. The unused, irrelevant and redundant attributes are removed from the dataset for its practical use. Attributes such as city, number of items, quantity, revenue generated, date, feedback, etc., can be considered as influencing attributes. The overall system architecture is given in Figure 5.2.

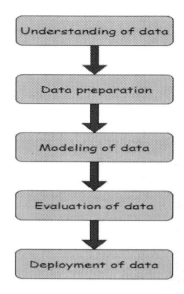

FIGURE 5.2 Stages in data mining.

5.2.2 Data Analysis

The nature of the data is analyzed during the data analysis phase. This step gives a clear understanding of the data we are working on. It involves data mining stages, such as the following.

5.2.3 Outlier Detection

This process is to perform the optimization of the model. Outlier detection helps analyze overall data to determine attribute values that are independent of the models. Its emphasisis on the quality check of the attributes. It is used to remove irrelevant or outlying data.

5.2.4 Training the Dataset

For any supervised learning algorithms to be used training, the dataset is a vital part. The more we train the model, the more accurate our results will be. Thus, the prediction of data can be quickly made by training the classifier model constructed.

5.2.5 Prediction

Forecasting the sales based on customer behavior is done using various machine learning (ML) algorithms. These ML algorithms are used to reduce manual intervention in the case of exhausting Big Data from e-commerce sites. The probability that the customer purchases a particular product is determined by predictive

Customer Behavior Prediction 83

models constructed. The predictive modeling approach is broadly classified into the individual level (i.e. customer product pair) and segment level approach (i.e. customer segments are formed using clustering) (Peker et al., 2018; Zhuang et al., 2019; Zhao et al., 2019). The classification and clustering of the sales dataset are done using ML algorithms. This survey conducts a review on neural network techniques and a few other machine learning algorithms which are used to forecast, namely random forest, support vector machine (SVM), decision tree, and logistic regression.

5.2.5.1 Random Forest

The collected dataset is preprocessed, then analyzed and important customer behavior attributes are extracted from the large dataset. To construct a forest of trees determining customer behavior, random forest is to be applied to the extracted features. Bagging and feature randomness is used to build individual tree, which is then classified into targeted customer behaviors. Then the prediction is made using the voting technique.

5.2.5.2 Support Vector Machine

SVM can be employed to perform regression and classification. A multidimensional hyperplane is used to perform classification. Neural network techniques were used to build SVM. The SVM algorithm identifies the hyperplane in an N—a number of features—and classifies the customer behavior distinctly based on the regularity of the customer visits to the e-commerce sites (Maheswari & Packia Amutha Priya, 2017).

5.2.5.3 Decision Tree

The decision tree classifies customer behavior by recursive partitioning. It can be employed to analyze multiple variables. It is used to assign a label to customer behavior based on the highest score.

5.2.5.4 Logistic Regression

Logistic Regression (LR) is a predictive analysis algorithm that is used for classification problems. LR is used to predict the probabilistic value of the categorical dependent value between 0 and 1. It is used to determine the most influential variable that can be in classification. The relationship between the customer behavior and willingness to buy a product can also be derived using LR methodology, which can be used to classify the customer behavior into segments to predict the next basket list (Peker et al., 2018).

5.2.5.5 Naive Bayes

Naive Bayes is a classification algorithm used to solve classification problems. It makes use of probability to make predictions. It is a probabilistic classifier that learns the pattern and compares the contents with the list of words to classify the most probable class label for each object. When multiple features are described in the dataset, naive Bayes algorithm is broadly used. It can be used to forecast whether a customer will purchase the product or not. Naive Bayes is used to derive accurate results for large datasets. It is used to analyze customer behavior to suggest offers based on the customer's interest (Arsalwad & Dhanawade, 2017).

5.2.5.6 K-Means Clustering

K-means algorithm is the most popular clustering algorithm that is used to partition the given data into distinctive non-overlapping subgroups iteratively until constant centroids are obtained. The iteration continues until we get distinct subgroups with constant centroids. K-means clustering is used to arrive at meaningful insights and recommendations to generate customer segments based on customer behavior (Vivek, 2018).

5.3 TECHNIQUES EMPLOYED

After the successive collection of sales data, preprocessing is done to extract their accurate information for further detection. Also supervised machine learning classification algorithms namely logistic regression, random forest, naïve Bayes's, SVM and decision tree are employed. The data mining technique is used for the identification of patterns in the collected data. Traditional ML algorithms, hybrid model approach and neural network approach are deployed. Finally, for the comparison between different ML algorithms that were used in our work, individual- and segment-level approach have been used.

5.3.1 Machine Learning Approaches

Machine learning is considered to be part of AI. It serves as an automated tool to find the hidden patterns in voluminous data. ML techniques can be broadly classified into four: namely supervised, unsupervised, semi-supervised and reinforcement learning.

The collection of e-commerce datasets from the internet is the first and foremost activity in sales and customer behavior. Then the preprocessing of data and outlier detection is done to determine the general information from the dataset and correlation between the attributes. To train the classifier, a training dataset is being used in the e-commerce dataset and the classification model is built using supervised machine learning classification algorithms, namely logistic regression, random forest, naive Bayes, SVM and decision tree. The classification model can also be built using the unsupervised algorithm such as K-means clustering, K-nearest neighbors whenever the initially labeled dataset is not obtained. Performance evaluation is done using the test data to determine the correctness of the classification.

To improve the prediction of sales, comprehensive predictive models of machine learning are used which also deal with Big Data. An analysis can be conducted to come up with the best-suited machine learning algorithm based on the evaluation of performance (Cheriyan et al., 2018; Piyush Anil et al., 2019). Machine learning is widely used to classify customers based on their purchasing behavior as regular, occasional, festival, offer or window shoppers based on their visits to the e-commerce sites. Then the frequency in which the product is purchased and which age group

Customer Behavior Prediction

of shoppers bought the product more can be analyzed using machine learning algorithms to predict the sales forecast (Maheswari & Packia Amutha Priya, 2017).

The relationship between customer behavior and willingness to buy a product can also be achieved using machine learning methodologies. ML algorithms can be used to classify customer behavior into segments to predict sales (Harsh Valecha et al., 2018; Lin et al., 2018; Xu et al., 2018). A personalized shopping list for each customer based on their behavior can be derived. This can be used to predict the customer's next basket shopping list using an individual-level and segment-based approach. This also employs different machine learning predictive algorithms (Peker et al., 2018; Maksim Korolev & Ruegg, 2015).

5.3.2 DATA MINING APPROACHES

A technique that excerpts a large volume of information from raw and vague random data and converts it into useful information is known as data mining (Yadav et al., 2017; Victor Haastrup Adeleye et al., 2014). It is used to extract useful hidden patterns from large datasets. The data mining technique is used to preprocess history data and present business levels to forecast future sales and customer behavior. Strategic decisions have to be made based on the analysis report generated by data mining to implement new policies that help to satisfy the customer better. Data mining can be described as AI techniques and mathematical statistics.

The data mining technique has the stages shown in Figure 5.3.

5.3.2.1 Data Preparation

The data preparation phase is the next forwarding phase to the data collection from e-commerce sources. Data preparation involves the preprocessing of data obtained from the e-commerce site. Data preprocessing such as data cleaning to remove the redundant and less useful attributes from the dataset, transformation of data, subset selection is done here.

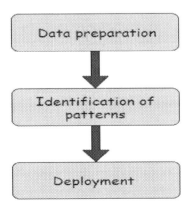

FIGURE 5.3 Data mining techniques.

5.3.2.2 Identification of Patterns

By considering and comparing various models, the best model is selected based on its performance. The preprocessed data has a high-quality dataset, which is used to extract client behavior patterns that are redundant or based on a particular feature. This customer pattern helps to plan out strategic decisions to balance the demand-driven supply chain.

The discovered pattern is then analyzed for formulating the standard behavior of the customer using the e-commerce site. The correlation regulation discovery is made to predict the next page of the client using the prefetching technique. Sequence pattern recognition is used to find the time sequence (i.e. which purchase pattern follows another). The directed tree model can be used to perceive the maximum forward path (MFP) and often visited the path by the customers (Yadav et al., 2017; Tonya Boone et al., 2019).

The clustering of the patterns is done usually using an unsupervised algorithm, K-means clustering. The K-means algorithm groups the customers into six main clusters based on the behavior and timing of when the people access the e-commerce site (Yadav et al., 2017; Jiang & Yu, 2008; He & Jiang 2017). Data mining proceeds to analyze the customer's basket to determine the pattern to predict the next basket of the customer (Victor Haastrup Adeleye et al., 2014; Wang et al., 2015).

Data mining is a critical process. Collecting and preparing the data for the extraction of specific information and identifying a particular pattern to analyze customer behavior from a large velocity of data is a crucial process. Therefore, many retailing companies purchase databases to order the data in the long run. GenMax, a new algorithm, was introduced to utilize a backtracking search to enumerate the maximal frequency itemset (MFI) for maximal pattern identification (Burdick et al., 2005; Karim et al., 2012).

The model can be designed using mining techniques to give personalized recommendations based on the predicted behavior of the customer (Rana Alaa El-Deen Ahmeda et al., 2015; Eric Michael et al., 2017). There is a rapid growth in the amount of data in the e-commerce field, as there are no geographical limitations in the e-commerce sites and the availability of products is humongous. Data analyst faces lots of issues in sales forecasting. The essential functions of sales prediction are the identification of valuable attributes, a new product in the market, fixing the price that suits globally and net sales realization (Sastry et al., 2014; Jain & Chandra, 2017).

5.3.2.3 Deployment

The deployment phase involves using the best model selected in the pattern recognition phase and employing the model on the data to predict or forecast the recommendation list. The selected model may be composed of ML, neural networks or any hybrid techniques.

5.3.3 HYBRID MODEL CONSTRUCTION APPROACH

Hybrid models are constructed by combining two or more machine learning methods or using other optimization techniques that enhance performance prediction.

Customer Behavior Prediction

Hybrid models are built by combining different grouping methods like bagging or boosting to make up a highly accurate classifier model.

Hybrid model construction can be adapted to construct a two-level approach by combining data mining predictive techniques over a stack. The model contains two layers of predictive algorithms to forecast sales. In (Kumari Punam et al., 2018), the first level consists of predictive algorithms such as linear regression, regression tree, cubist, support vector regression and K-nearest neighbor is an ensemble; and the second level is built using the linear regression, support vector regression and cubist algorithms. The autoregressive integrated moving average model (ARIMA) is used in analyzing the time series of any dataset to predict future sales. In (Rankothge Gishan Hiranya Pemathilake et al., 2018), a hybrid model is built using an integration of the ARIMA and RNN module to determine the correlation between the input dataset attributes.

5.3.4 NEURAL NETWORK APPROACH

The neural network approach is an analytical technique that is designed based on the neurological functioning of the brain to untangle and provide a solution for a complex problem. Here the learning process is done using the cognitive system. New predictions can be made based on the observation of the existing data. The network is designed to match the investigation problem that involves many trial-and-error methods. Artificial intelligence is employed to find the best network. This network architecture is then trained by applying various numbers of inputs and adjusting the weight of the network iteratively to predict customer behavior (Cui et al., 2018; Kuen-Han Tsai et al., 2017; Jia et al., 2017) optimally.

5.3.5 INDIVIDUAL- AND SEGMENT-LEVEL APPROACHES

The individual-level approach uses the past behavior of an individual customer product purchase and transaction to build a model for that individual customer and the product recommendation. The segment-level approach adapts the aggregation technique to predict the purchased product of a targeted group of customers (Peker et al., 2017; Peker et al., 2018). The predictive model is built using ML algorithms suited.

5.4 PERFORMANCE EVALUATION AND RESULT

For classification models constructed, the terms true positives (TP), false positives (FP), real negative and false negative are defined based on the observations.

- **Recall:** The probability of total relevant instances that are retrieved.

$$Recall = \frac{TP}{TP + FN} \tag{1}$$

- **Precision:** The probability of relevant instances that are obtained among the total retrieved instances.

$$Precision = \frac{TP}{TP + FP} \tag{2}$$

- **F-measure:** A measure that integrates recall value and precision value. It is said to be the harmonic mean of recall and precision values.

$$F - measure = 2 \times \frac{Precision \times Recall}{Precision + Recall} \tag{3}$$

- **mRHR:** A metric used to evaluate the ranking of the recommended product by the model in case there exist multiple preferences for the customer.

$$mRHR = \frac{1}{preferences} \sum_{i=1}^{N} \frac{hit}{rank} \tag{4}$$

Where hit is if the customer prefers the product predicted by the model. Rank is the position of the preferred product.

To achieve experimental results, this survey analyzes the customer behavior of data using various techniques that include machine learning such as logical regression, naive Bayes, SVM, decision table, random Forest and neural networks using a predictive model approach to segment customers and forecast the shopping menu. The comparison of the ML algorithms is made using performance evaluation factors such as F-measure and mRHR. Table 5.1 shows that SVM gives the best result with 0.345 of F-measure and 0.120 of mRHR in an individual-level approach. LR exhibited the best result in the segment-level approach, with an F-measure of 0.355 and

TABLE 5.1

Comparison of Customer Behavior Modeling Using Various ML Algorithms

Individual-Level Approach		
ML Algorithm	F-measure Value	mRHR Value
Logistic regression	0.328	0.100
Decision tree	0.329	0.107
Random forest	0.337	0.112
Support vector machine	0.345	0.120
Neural network	0.314	0.102
Segment-Level Approach		
ML Algorithm	F-measure Value	mRHR Value
Logistic regression	0.355	0.124
Decision tree	0.33	0.112
Random forest	0.315	0.109
Support vector machine	0.322	0.113
Neural network	0.293	0.092

Customer Behavior Prediction

an mRHR of 0.124. LR of the segment level approach recorded the overall best performance. Figure 5.4, Figure 5.5, Figure 5.6 and Figure 5.7 compare the F-measure and mRHR values of all ML algorithms using predictive models constructed. The survey result indicates that the performance of the model constructed depends on

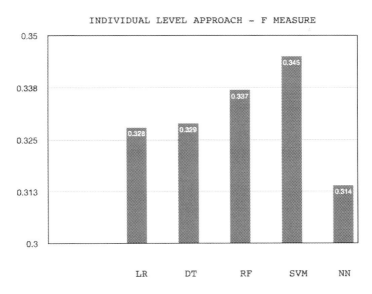

FIGURE 5.4 Comparison of F-measure for all ML algorithms using the individual-level approach.

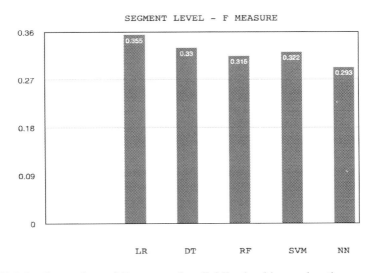

FIGURE 5.5 Comparison of F-measure for all ML algorithms using the segment-level approach.

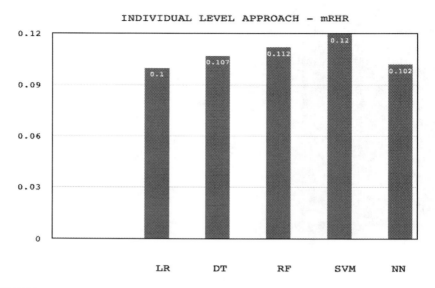

FIGURE 5.6 Comparison of mRHR for all ML algorithms using the individual-level approach.

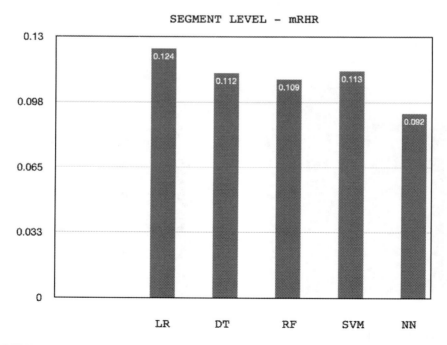

FIGURE 5.7 Comparison of mRHR for all ML algorithms using the segment-level approach.

Customer Behavior Prediction 91

the algorithm used. Thus, it is essential to experiment with different ML algorithms before the construction of the model to analyze customer behavior.

5.5 CONCLUSION

Accurate prediction on the customer behavior for suggesting a shopping list or forecasting the next basket of the customer is very critical to provide efficient, customized service to the customer. The prudent way to maximize revenue is to improve customer satisfaction. This survey discusses various approaches such as machine learning, data mining and neural network employed by the data analyst to analyze customer behavior. A comparison of the machine learning algorithms is made using performance evaluation measures, and the logistic regression is selected as the best predictive model which accurately predicts the interests of consumers.

REFERENCES

Arsalwad GP and Dhanawade AA (2017), "Analysis of e-customers behavior using Naïve Bayes algorithm", Available: www.ijircce.com/upload/2017/april/169_ANALYSIS.pdf.

Bowen T, et al. (2020), "Forecasting method of E-commerce cargo sales based on ARIMA-BP model", *2020 IEEE International Conference on Artificial Intelligence and Computer Applications (ICAICA)*, Dalian, China (pp. 27–29).

Burdick D, et al. (2005), "An efficient algorithm for mining maximal frequent item sets", *Data Mining and Knowledge Discovery* (pp. 223–242).

Cheriyan S, et al. (2018), "Intelligent sales prediction using machine learning techniques", *International Conference on Computing, Electronics & Communications Engineering (access)*", Southend (pp. 16–17).

Cui Y, et al. (2018), "Modelling customer online behaviours with neural networks: applications to conversion prediction and advertising retargeting", Available: https://arxiv.org/abs/1804.07669.

Eric Michael S, et al. (2017), "Forecasting customer behaviour in constrained E-commerce platforms", *International Conference of Pattern Recognition Systems (ICPRS 2017)*, Madrid, Spain (pp. 11–13).

Harsh Valecha, et al. (2018), "Prediction of consumer behaviour using Random Forest algorithm", *5th IEEE Uttar Pradesh Section International Conference on Electrical, Electronics and Computer Engineering (UPCON)*, Gorakhpur, India (pp. 2–4).

He J and Jiang W (2017), "Understanding users' coupon usage behaviours in E-commerce environments", *IEEE International Symposium on Parallel and Distributed Processing with Applications and IEEE International Conference on Ubiquitous Computing and Communications (ISPA/IUCC)*, Guangzhou, China (pp. 12–15).

Jain A and Chandra S (2017), "Sales forecasting for retail chains", Available: www.semanticscholar.org/paper/Sales-Forecasting-for-Retail-Chains-Jain-Menon/76a244f4da1d29170a9f91d381a5e12dc7ad2c0f

Jia R, et al. (2017), "E-commerce purchase prediction approach by user behavior data", *International Conference on Computer, Information and Telecommunication Systems (CITS)*, Dalian, China (pp. 21–23).

Jiang Y and Yu S (2008), "Mining E-commerce data to analyse the target customer behavior", *First International Workshop on Knowledge Discovery and Data Mining* (pp. 406–409).

Karim M, et al. (2012), "Mining E-Shopper's purchase rules by using maximal frequent patterns: An e-commerce perspective", *International Conference of Information Science and Applications (ICISA)*, Suwon, South Korea (pp. 23–25).

Kuen-Han Tsai, et al. (2017), "Multi-source learning for sales prediction", *2017 Conference on Technologies and Applications of Artificial Intelligence (TAAI)*, Taipei, Taiwan (pp. 1–3).

Kumari Punam, et al. (2018), "A two-level statistical model for Big Mart sales prediction", *International Conference on Computing, Power and Communication Technologies (GUCON)*, Uttar Pradesh, India (pp. 28–29).

Li X, et al. (2020), "Characterizing social marketing behavior of E-commerce Celebrities and predicting their value", *IEEE INFOCOM 2020—IEEE Conference on Computer Communications Workshops (INFOCOM WKSHPS)*, Toronto, Canada (pp. 6–9).

Lin W, et al. (2018), "Social media brand engagement as a proxy for E-commerce activities: A case study of Sina Weibo and JD", *IEEE/WIC/ACM International Conference on Web Intelligence (WI)*, Santiago, Chile (pp. 3–6).

Maheswari K and Packia Amutha Priya P (2017), "Predicting customer behaviour in online shopping using SVM classifier", *IEEE International Conference on Intelligent Techniques in Control, Optimization and Signal Processing (INCOS)*, Srivilliputhur, India (pp. 23–25).

Maksim Korolev and Ruegg K. (2015), "Gradient boosted trees to predict store sales", Available: http://cs229.stanford.edu/proj2015/193_report.pdf.

Peker S, et al. (2017), "A hybrid approach for predicting customers' individual purchase behavior", *Kybernetes* (46(12), pp. 1–22).

Peker S, et al. (2018), "An empirical comparison of customer behaviour modelling approach for shopping list prediction", *International Convention on Information and Communication Technology, Electronics and Microelectronics (MIPRO)*, Opatija, Croatia (pp. 21–25).

Piyush Anil B, et al. (2019), "Designing a sales prediction model in tourism industry and hotel recommendation based on hybrid recommendation", *Third International Conference on Computing Methodologies and Communication (ICCMC 2019)*, Erode, India (pp. 27–29).

Rai S, et al. (2019), "Demand prediction for e-commerce advertisements: A comparative study using state-of-the-art machine learning methods", *International Conference on Computing, Communication and Networking Technologies (ICCCNT)* (pp. 6–8), Kanpur, India.

Rana Alaa El-Deen Ahmeda, et al. (2015), "Performance study of classification algorithms for consumer online shopping attitudes and behavior using data mining", *Fifth International Conference on Communication Systems and Network Technologies*, Gwalior, India (pp. 4–6).

Rankothge Gishan Hiranya Pemathilake, et al. (2018), "Sales forecasting based on AutoRegressive Integrated Moving Average and Recurrent Neural Network hybrid model", *International Conference on Natural Computation, Fuzzy Systems and Knowledge Discovery (ICNC-FSKD)*, Huangshan, China (pp. 28–30).

Sastry, SH et al. (2014), "Analysis and prediction of sales data in SAP-ERP system using clustering algorithms", Available: arXiv preprint arXiv (p. 1312.2678).

Singh B, et al. (2020), "Sales forecast for Amazon sales with time series modeling", *First International Conference on Power, Control and Computing Technologies (ICPC2T)*, Raipur, India (pp. 3–5).

Tonya Boone, et al. (2019), "Forecasting sales in the supply chain: Consumer analytics in the big data era", *International Journal of Forecasting* (35(1), pp. 170–180).

Victor Haastrup Adeleye, et al. (2014), "Customer behaviour analytics and data mining", *American Journal of Computation, Communication and Control* (pp. 66–74).

Vivek S (2018), "Clustering algorithms for customer segmentation", Available: https://towardsdatascience.com/clustering-algorithms-for-customer-segmentation-af637c6830ac.

Wang P, et al. (2015), "Learning hierarchical representation model for next basket recommendation", Available: www.bigdatalab.ac.cn/~junxu/publications/SIGIR2015_NextBasket Rec.pdf.

Xu D, et al. (2018), "Repurchase prediction based on ensemble learning", *IEEE SmartWorld, Ubiquitous Intelligence & Computing, Advanced & Trusted Computing, Scalable Computing & Communications, Cloud & Big Data Computing, Internet of People and Smart City Innovations*, Guangzhou, China (pp. 8–12).

Yadav, M. P., et al. (2017), "Mining the customer behavior using web usage mining in e-commerce", Available: https://nevonprojects.com/customer-behavior-prediction-using-web-usage-mining/

Zhao B, et al. (2019), "Loyal consumers or one-time deal hunters: Repeat buyer prediction for E-commerce", *International Conference on Data Mining Workshops (ICDMW)*, Beijing, China (pp. 8–11).

Zhuang Q, et al. (2019), "A neural network model for China B2C ecommerce sales forecast based on promotional factors and historical data", *International Conference on Economic Management and Model Engineering (ICEMME)*, Malacca, Malaysia (pp. 6–8).

6 The Impact of Artificial Intelligence on Global Business Practices

Bhakti Parashar and Geeta Rana

CONTENTS

6.1 Introduction 95
    6.1.1 Evolution and History of Industrial Revolutions 97
    6.1.2 Fourth Industrial Revolution 98
    6.1.3 Artificial Intelligence, Machine Learning and Deep Learning 100
6.2 Need for Artificial Intelligence in Global Business 102
6.3 Future Impact of Artificial Intelligence on Global Business Practices 103
    6.3.1 Impact of Artificial Intelligence in Various Sectors 104
6.4 Inferences 108
6.5 Conclusion 109
References 110

6.1 INTRODUCTION

The first person who proposed artificial intelligence was John McCarthy in the year 1956 on this topic in his first academic conference. His idea that machines operating like humans was one of the centers of the scientist's mind, and a point of discussion about whether machines will have the equal ability to learn and think was introduced by the mathematician Alan Turing. He was capable of putting his hypothesis and questions into actions with the use of testing whether machines can think. After some more testing, which was later known as Turing test, came with the result that it is possible to empower machines to learn and think just like humans. The Turing test was used as a pragmatic approach to determining whether or not machines can respond like humans (Smith et al., 2017).

AI has been defined as that activity which is devoted to making machines intelligent and that intelligence can be that quality that enables an entity to perform appropriately and with foresight in its environment (Nilsson, 2010). AI is not restricted but is applicable to many fields like linguistics, psychology, biology, cognitive sciences, neuroscience, philosophy, mathematics, logic, engineering and computer sciences; it can contribute in almost every field to achieve more success. Regardless of the particular approach, artificial intelligence has been combined from the start by its engagement with the possibility of mechanizing intelligence (Turing, 1950).

DOI: 10.1201/9781003145011-6

In terms of potential, AI has the potential to imitate the human brain, as well, which makes it different and unique as compared to other technologies as it can understand and solve the problems that generally need human intelligence. More often, AI includes natural language and processing, visual perception and pattern recognition, and decision-making, as well and this processes in combination give much more potential to AI in multiple disciplines and many economic sectors across the globe. Furthermore, AI may help to address persistent development challenges like lack of infrastructural facilities or an underdeveloped health care and financial sector which may lead to deprived individuals. In its most basic form, despite its revolutionary potential, AI has existed for decades. In its first generation, AI-equipped computers were able to play chess, solve puzzles and perform other related straightforward tasks (Mou, 2019).

At present, we can also say that AI has enormous prospective to augment human intelligence and to radically modify the way a person accesses goods and services, collects information, makes goods and interacts. In emerging economies, AI offers an opportunity to reduce the costs and barriers to entry to a business and gives innovative business models that can jump from tradition solutions to advanced ones and reach the underserved. Technology-based solutions are absolutely important for the economic development of most of the nations, for the removal of poverty and boosting shared wealth may become dependent on connecting with the AI power. Although emerging markets are already using basic AI technologies to resolve critical development challenges, a lot more can be done, and private sector solutions will be critical to mapping the new business models, developing new means of delivering services and increasing local markets competitiveness. All these solutions need innovative approaches to extend prospects and alleviate risks associated with the new technology (Strusani & Houngbonon, 2019).

Every country grows in stages. With the eruption of the dot-com boom, innovative trade ideas and prototypes have arisen; merchandise has evolved virtually and e-commerce has also been expanded, along with other financial services in the online domain with digital payments such as Paytm, Google Pay, online payments and banking; by augmenting processes and procedures, these services have fastened the transactions and there will be considerable changes occurring in the cycle. This substantial change is carried out by the subsequent generation of industrialization with the augmentation of technologies, resulting in upliftment of job opportunities (Pavaloiu, 2016; Rana, 2010). With the rapid advancement of artificial technologies, businesses in emerging economies have begun to use AI applications to boost productivity and identify new business opportunities. AI applications are gaining in popularity in the world's fastest-developing countries, such as China. Machine learning and deep learning have made tremendous progress due to their rapid expansion and improvement (Bughin et al., 2017). It can be considered as the new source of energy toward trade strategies and models, which can be used in identifying various viable business situations, forecasting the novel conditions and yielding brilliant and swift business decisions which can raise the revenue of a business (Coats, 1987).

In simple words, AI can be defined as the ability of machines to perform different tasks just like human do with their brains; this has attracted the American

Impact of AI on Global Business Practices 97

public's imagination in ways both good and bad. Users of the latest technology like computers and smartphones are already embracing the use of AI technologies in the form of internet searches, speech recognition, Facebook image tagging, etc. People are now expecting digital special effects in movies and the three-dimensional reconstructions of key plays in realistic video which is only possible by AI technologies.

6.1.1 Evolution and History of Industrial Revolutions

The industrial revolution is primarily known as an era in which work started to be accomplished more and more by machines in industrial units rather than in homes. More advancement in science and automation have constantly encouraged world and industrial progress in international business, facilitating us to provide more definite as well as clear meanings to the term in recent years (Belvedere et al., 2013). It is a concept and development that has changed our economy and society. The world development term may give the impression to point out some tardiness in this perspective of revolution, which truly indicates rapid and fundamental changes—undoubtedly, that key changes have happened within a relatively short period. Industrial development led to the replacement of small-scale factories, workshops and craft studios. The new beginning of industries was first determined with textile and pottery factories, and entered into the new infrastructure of canals and railway lines, which enabled the efficient distribution. This was the actual conversion of industrious to industrial, and their start of boom for both started (Van et al., 2014).

The industrial revolution has always in the past been linked with the advancement of technological competencies. Persuasion of improved proficiency and yield makes the industrial sector one of the major players for the period of technical and industrial transformation and conception. Today, advancement and transformation is meant by the Fourth Industrial Revolution to facilitate an upgrading in technologies that could help us to fulfill additional requirements which are presently assimilated into dispersed cyber-physical systems (CPS) in ingenious industrial units (Saturno et al., 2017).

There is no universal agreement these days on what constitutes an industrial revolution (Maynard, 2015), but according to the National Academy of Science and Engineering (2013), the general categorization of stages that have been defined from the perspective of industrial revolutions are as shown in Figure 6.1.

The first Industrial Revolution is considered as the most significant in terms of civilization which begins with using water and steam–powered motorized engineering facilities at the end of the eighteenth century. The twentieth century started with the Second Industrial Revolution, the utilization of electrically powered mass production machinery with the use of division of labor. Later promoting automated manufacturing, the Third Industrial Revolution started in the mid-1970s by disseminate electronics and information technology (IT) in manufacturing units. The Fourth Industrial Revolution began with the use of artificial intelligence machine learning, the Internet of Things (IoT), and much more, as the IoT and cyber-physical systems have attracted and caught the attention of researchers (Atzori et al., 2010).

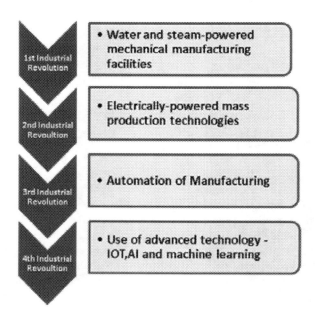

FIGURE 6.1 Global development of industrial revolutions.

6.1.2 FOURTH INDUSTRIAL REVOLUTION

Industry 4.0, or the Fourth Industrial Revolution, gives further advancement in technology in the form of the IoT, data science, Big Data, cloud computing, blockchain and artificial intelligence—all of these technologies are changing the way we live, work and are entertained. Moreover, these technologies can contribute to developing hyperautomation and hyperconnectivity (Schwab, 2017; Rana & Sharma, 2019; Bloem et al., 2014; Klosters, 2016; Park, 2017). It fuels advancements in science and technology, with IoT and related technologies serving as pillars of CPS, in which machine intelligence is used as agents to augment production lines.

Agility, intelligence and networking are all aspects of improvement that transcend organizational and national boundaries (Liao et al., 2017). Until now, Industry 4.0 is not defined and restricted to computer technology, IoT and other networks to improve production. This Fourth Industrial Revolution takes computing to a new level, toward a cyber-physical world where augmented reality, artificial, data-driven decision-making autonomous systems and the cloud are changing major manufacturing industry standards. Higher skilled employees working collaboratively with robots in a horizontal organization are some of the predicted impacts. CPS under Industry 4.0 plays an important role in the advancement and fast development of communication, and we can make it more common in the society when it connects with persons, but it is essential to confirm that CPS performs firmly and has assured tolerance once utilized through artificial intelligence (Mosterman & Zander, 2015). CPS is moreover the base to build the IoT that could be joined to develop as an

Impact of AI on Global Business Practices 99

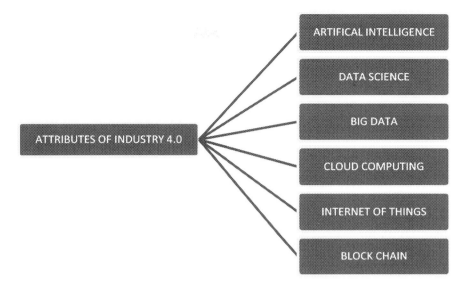

FIGURE 6.2 Major attributes of Industry 4.0.

internet of services, as well. Thus, an organization will find it easier to create worldwide networks which can be linked with the warehousing arrangements, equipment and manufacturing facilities of CPS in the future (He, 2016).

Fourth Industrial Revolution or Industry 4.0 has many subparts and attributes where artificial is one of the most important element.

Artificial intelligence: Advanced AI can be called rational when process control includes self-directed modifications to the operating system which can be performed through simulation. As per the latest revolution, the simulation modeling pattern facilitates modeling of the production process and other processes with the use of computer-generated manufacturing concepts. Simulations are completed with the concurrent data to symbolize the real world in a simulation model that comprises human beings, goods and machinery. Hence, machinists can augment the machine settings in a computer-generated simulated situation earlier applying in the real world, and this leads to reductions in setting machine setup times and increases the quality (Rodič, 2017).

Data science: Data sciences are a combination of statistics, information, computing, communication, sociology and management which contain social aspects, and science which acts based on data, environment and the so-called data-to-knowledge-to-wisdom thinking (Cao, 2017).

Big data: Data analytics has now become the recent topic in Industry 4.0, as it is very useful for predictive manufacturing and works in the direction of industrialized technology progress with advanced internet use. An enormous amount of data can be and taken with Big Data and analytics on daily basis. Big Data analysis can be

100 Reinventing Processes Through AI

used to obtain additional value when the present technique becomes more mature to manage Big Data. Thus, Big Data is the utilization of digital technology to do analysis (Sharma et al., 2021b). There can be four dimensions of Big Data i.e., volume, variety, value and velocity (Witkowski, 2017).

Cloud computing: Cloud computing is a relatively very new system logic which gives big storage capability for the operator and it requires less expense to organizations and individuals to access these resources. Eventually, the performance of technologies will improve, but the functionality of machine data will still be kept into the cloud storage system and will create a more data-driven overall production system. Restrictions on the organization can be curtailed if additional data sharing will be allowed through locations for associated manufacturing industries in this industrial revolution. Undoubtedly, cloud computing is attracting many organizations in the process of building data system (Xu, 2012).

Internet of Things (IoT): Nowadays, Industry 4.0 is experiencing a different combination of the present IoT and the manufacturing Industry. The outcome of this Industry 4.0 is a blending of the IoT and the internet of services in the production methods (Prakash et al., 2021; Kagermann et al., 2013). In general, IoT avails modern connectivity of systems, facilities and tangible goods, and enables object-to-object communication and data. IoT can be attained with the control and mechanization of aspects like heating, lighting, machining and remote monitoring in various industries (Zhong et al., 2017).

Blockchain: Blockchain can be considered as an advance technique that improves customer services, initiatives and end-to-end value, and raises the proficiency of procedures (Agarwal, 2018). It also allows unreliable and new investors to generate shared and safe data records; when an exchange of key data and products is needed, blockchain technology accelerates the businesses settlements, makes the procedure streamlined, decreases unwanted data and reduces production costs (Wasserman, 2016).

6.1.3 Artificial Intelligence, Machine Learning and Deep Learning

Artificial Intelligence is a very complex thing that has several concepts and different definitions. A very basic artificial intelligence program or automation has a very high level of risk as compared to a self-driven program, although when it comes to the paybacks to the business, artificial intelligence overcomes the risk. AI is already in use in business and probably continues to reach a height. It will be more prevalent in the business process which will be needed to be adjusted by society as well. It has been observed that all the forecasts for the future development of technology such as artificial intelligence are self-assured because they are diverse (Armstrong et al., 2014).

There are so many definitions given by various authors on artificial intelligence, machine learning and deep learning, and as per them, the involvement of all these three in human activities is unavoidable and has continued to raise concerns about their excellent payback to human and society as compared to taking over human jobs, as they have developed many processes starting from finance, healthcare services, e-commerce, engineering services, and many more.

Impact of AI on Global Business Practices 101

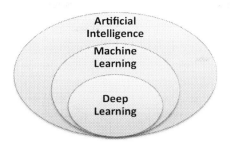

FIGURE 6.3 The relationship between artificial intelligence, machine learning and deep learning.

Although there are many subfields of AI, as per the research and studies, deep learning is a subset of machine learning—and further, machine learning is a subset of artificial intelligence. Although there are so many subfields of AI like machine learning, language processing, image processing and data mining, machine learning developed as a branch of artificial intelligence and deep learning as one of the parts of machine learning. With deep learning and machine learning, use of artificial intelligence has spread to so many fields such as medical science, smart manufacturing, agriculture, pharmacology, games, business, and so on (Cioffi et al., 2019).

Machine learning: Deep learning is a member of one of the families of machine learning algorithms, which is inspired by the biological process of neural networks that control many applications and gives benefits over conventional machine learning algorithms (Goodfellow et al., 2016). Deep learning uses its capability to produce faster and give more accurate results. It endeavors to model high-level abstraction in data based on a set algorithm (Deng & Yu, 2014). It is more suitable when the amount of training data is increased, with the increase in the software and hardware infrastructure development the development of the deep learning model has also been increased (Nisbet et al., 2018).

Deep learning: Learning can be defined as one of the processes of transformation and improvement in the behaviors through identifying new information in time—and when this learning is performed by the machines, it is called machine learning (Sırmaçek, 2007). The main objective of machine learning is to create a model which can train itself for improvement, perceive complex patterns and find solutions to new problems with the help of previous data (Tantuğ & Türkmenoğlu, 2015).

Machine learning and deep learning as a part of artificial intelligence are used in social network analysis where unsupervised machine learning algorithms can automatically recognize the networks within a user circle in Google or Facebook or it can determine the maximum number of emails sent to a specific person and classify into collective groups even if you want to know which groups of persons that all know each other, that can also be possible with AI. Nowadays most companies have a big database of their customers, here unsupervised machine learning algorithm can be used to look into the customer dataset and automatically discover market segmentation and set a group of customers into various market segments so the organization can automatically sell their products or market various market segments altogether more efficiently (Haider et al., 2012).

6.2 NEED FOR ARTIFICIAL INTELLIGENCE IN GLOBAL BUSINESS

Artificial intelligence is not only used in technological development but also applicable in marketing and business like automation in business methods, getting visions from the data or attracting consumers and engaging workers (Davenport & Ronaki, 2018). In emerging economies, artificial intelligence (AI) provides a technical solution to economic challenges faced not only by businesses and governments, but also by individuals at the bottom of the economic pyramid. Integration of data from a variety of sources, such as social media, websites, and conventional channels, will aid in the development of a firm's data management platform, successful business strategies, and the reduction of barriers to doing business (Arora & Rahman, 2017).

Artificial intelligence can be linked to frameworks derived from traditional techniques and models in today's world, with a focus on emerging economies. The fundamental concepts of microeconomic theory can be reassessed scientifically for the manifestation of new devices for making and augmenting rational decision-making, such as for the manifestation of new devices for making and augmenting rational decision-making. (Parkes & Wellman, 2015). The introduction of an asset-based model which can be fit for the employment shift was proposed as a radical intervention by economist McAfee whose focus was the combined approach on the basis of socialism, allocating prospective government assets evenly. An alternative basic earnings have put forward to build a new working paradigm. Thus, there is no need for people to involve in arduous administrative tiresome jobs and shall be free from dragger and toil, consequently, one's own value can be reconsider. With the practice of AI base technology, people may start working with pleasure, not with pressure to get pecuniary consistency or there can be one option to remain idle, as everything will be very systematic and time done (McAfee, 2013).

Artificial intelligence has that much potential to rebuild skill demands, career prospects and the systematic distribution of workers and employees between industries and professionals. Although people involved in researches and policymaking are not sufficiently equipped for labor trend forecasts resulting from specific cognitive technologies like AI. Technology is meant to perform a typical or specific type of task that alters the demand for specific skills at the workplace (Sharma & Rana, 2021). The result of such alteration to skill demands can be drawn out all over the economy, inducing occupational skill requirements, career mobility and social well-being and workers will have some social identity.

There may find some hurdles in the specific pathways to identify such alternatives which can be constrained by abrasive historical data and limited tools for the modeling resilience, which can be overcome by prioritizing data collection in-depth, responsive to actual dynamics in the labor market, and respective regional inconsistency, better accessibility to unstructured skills data from the résumé of the candidates, job postings with latest indicators for technological changes like the use of patented data and models for inter-region and intra-region labor dependencies will facilitate a new and promising technique for better understanding and future forecasting of work. The improved technique of this data collection will help us to use new data-driven tools, adding machine learning application and systematic modeling as well which can more accurately reflect the complexity of the whole labor system. Now new data will lead to a new kind of research that enhances our learning and understanding of the impact of technology and artificial intelligence on the modernized labor market (Frank et al., 2019).

Artificial intelligence has reached that place where it can be used to take the practical world to financial decisions, chat with the public, play games against humans and work collectively with them. There is only an AI-driven system or intelligent agents (IA) behind these real world–based applications, which interacts with the environment in a repetitive manner of sense, think and act. It considers data from the environment to make an informed decision based on the input data and previous experience, and at last performs an action affecting the environment. It can work like a machine in industries, home robots or as self-driven cars; it can also work as a software agent like chatbots, recommendation systems, and many more. AI algorithms take the information in the form of images, sound, text, videos, etc., analyze it and deliver AI-powered solutions (Soni et al., 2019).

AI not only gives reliability but also provides cost-effectiveness and solves complex problems and helps to come to an accurate decision. Similarly, it controls the missing data will come; that is why it is useful in many fields of business and engineering. Reinforcement learning—which is based on testing success and failure in the real world to increase the reliability of applications—is one of the best tools used in AI, although AI also has its limitations with competency and functionality (Sadek & Chowdhury, 2012).

6.3 FUTURE IMPACT OF ARTIFICIAL INTELLIGENCE ON GLOBAL BUSINESS PRACTICES

With the development of AI, its practice has also increased in several ways in international business, which has changed the trade effect at the macroeconomic level; for example, as AI increases productivity and growth, it will lead to an increase in economic growth and gives new opportunities at a global level. Currently, the productivity growth rate at the global level is much less for various reasons (Remes et al., 2018)—the first reason is that countries take time to integrate and make effective use of new technologies such as AI, which is a relatively complicated one with economy-wide impacts (Brynjolfsson et al., 2017).

AI is likely to affect not only the type but also the quality of economic growth with international trade implications; it will fasten the transition toward service economies. Similarly, AI will also aid international trade in terms of expansion of automation and speed up job losses for the low-skilled and blue-collar workers in the manufacturing sector, as well (Arnet et al., 2016).

6.3.1 Impact of Artificial Intelligence in Various Sectors

AI will undoubtedly grow in the coming years, eventually becoming commonplace, and technology has made it possible for computers to use large quantities of data in decision-making processes or to learn on their own. Technology giants like Google, Apple and Facebook are already putting money, effort and time into AI integration. These businesses have also taken steps to educate the general public about AI's capabilities and limits. But this automation in business will result in substituting many fields; for example, jobs and skills will remain to contribute to the shortage of employment, as artificial intelligence is a compliment to some extent to labor and will not substitute for them entirely. Artificial intelligence is already being used in areas such as law, political science, policy and economics; it will soon be used in areas such as warfare, autonomous transportation, education and space exploration, to name a few.

Let's take a look at some of the potential and likely future AI applications that will undoubtedly improve people's lives. Under the influence of AI, the transportation, healthcare, logistics, finance and industrial manufacturing industries, among many others, would undergo massive transformations, allowing them to become more efficient, cost-effective and, most importantly, provide better services. Subsequently, artificial intelligence will have positive impacts on different sectors which somehow directly and indirectly affect the global business environment and are shown in Figure 6.4.

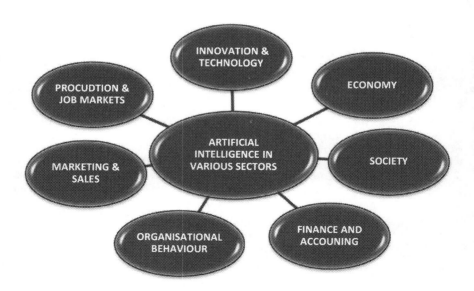

FIGURE 6.4 Impact of artificial intelligence in various sectors.

AI in innovation and technology: With the adoption of AI, the world is becoming smarter and more innovative. Use of Google Maps for route and traffic mapping, price determination of rides by Uber and Lyft, tagging a friend and get suggestions on Facebook, filtering spam emails, online shopping and recommendations and detection of disease are all some of the examples of AI technological innovations which have simplified our lives. The unbelievable internet speed with which AI is entering into every sector at the global level is pressuring organizations to adopt AI, which is also compelling business, policymakers, innovators, entrepreneurs and investigators to adopt AI for the development of new strategies and generation of new sources of business value (Soni et al., 2019).

Alibaba is the biggest e-commerce platform in the world, selling more than Amazon and eBay combined. Artificial intelligence (AI) is used to predict what consumers may want to buy in Alibaba's daily operations. The company uses natural language processing to create product descriptions for the website. Alibaba also employs artificial intelligence in other ways. Alibaba's City Brain project, which aims to develop smart cities, is another example of artificial intelligence in action. By tracking every car in the city, the project uses AI algorithms to help alleviate traffic congestion. Alibaba's cloud computing division, Alibaba Cloud, is also using artificial intelligence to help farmers monitor crops in order to increase yield and cut costs.

AI in the economy: By increasing efficiency, business process automation, financial solutions and government services, AI-based technologies can help expand the market and create more opportunities. Inequality and poverty can be reduced by empowering AI in the public and private sectors to collaborate on leapfrogging solutions in emerging and global markets (Andrews et al., 2019). Technological upgrades in societal development, improvement in production, marketing and finance are all signs of economic development, and artificial intelligence is complimenting these sectors to achieve better lives. Singapore is the country using AI technology the most. With the aim of becoming the global hub of AI technologies, Singapore has chosen five major AI-driven development projects that resolve national challenges while taking into account the effects on society and the economy.

AI in society: AI has its impact on humans and society as well, where its involvement gives a new meaning of work as self-worth, people will start work for pleasure instead of compulsion, and stability and security at workplace will be ensured. Although machines are not creative, nor have empathy, compassion or emotional intelligence, working with such machines will increase their emotional quotients which will change the outlook and attitude toward life which will lead to societal development (Pavaloiu, 2016). In recent years, AI has gotten a lot of buzz and has become more accessible to us as a result of internet-based innovation. It has risen to the forefront of many current debates as a result of these advances, as well as interest in the technology's potential socioeconomic and ethical consequences, and governments are attempting to comprehend the implications of technology for their people. In Singapore, educational institutions have placed a greater focus on fields such as mathematics, statistics, computer science and information technology in order for students to comprehend the importance and trends of data science, machine learning and artificial intelligence. Apart from their daily curriculum, business and management organizations are now focused on data patterns; it can be used by banks to determine whether there is unusual activity on an account. The algorithm could detect unusual activity, such as foreign transactions. Thus, the application of artificial intelligence in society will enable full-fledged digitization for faster development without corruption.

AI in finance and accounting: AI has also entered in finance through its digital footprint. AI application in finance and accounting includes recording of consumers' habits on the basis of their buying habits and frequency; their different actives related to finance and accounting, like their income and how they are managing it; and their financial investments and plans. With the involvement of AI in finance and business, the chances of fraud will be minimized; cost-saving specifically in businesses and AI can be organized in such a manner so that one can minimize risk and maximize profit more proficiently with increased security. Financial services will benefit from digital employees' 24/7 support, and our solution can be tailored to meet the needs of other departments. Process automation, case handling, bot operation, customer validation and other AI solutions are being developed. Businesses can use digital employees to scale their operations, increase service efficiency and cut costs all at the same time. Finance and accounting can be considered as two different concepts, although there is a lot of integration concerning the application of artificial intelligence to separate them. Both finance and accounting controls the business by analyzing performance, ensuring legal adherence, making budgets and developing strategies (Fredman, 2018).

The major benefit of AI is that it can collect a large amount of data and create information in a simplified way. Financial institutions have started adopting and in the future will adopt more of these automated systems to identify and minimize human errors, where resource-intensive, repetitive tasks are very suitable to be accomplished by artificial intelligence programs. In their finance and accounting systems, companies like Accenture use automation, machine learning and adaptive intelligence.

AI in organizational behavior: The ability to collect and process data is constantly improving, opening up new possibilities for reshaping how companies operate. Organizations should use AI to discover patterns and draw insights from data in order to make better decisions. Ethics and values of an organization will be minimized once the AI technology will be applied, as it will provide more transparency and less conflict and dishonesty in the workplace. The way of working of subordinates for the fulfillment of tasks and accomplishment of goals will go to another level, where the performance leads to employee's productivity, training and strategic management. In the presence of AI, an organization will be affected in two ways. First, reassessment of the distribution of work will be done with the influence of the new technology at the hierarchy level and changes required to be done by the replacement of the human workforce with AI, wherever it will be needed. Second, knowledge of the people and their perceptions of losing their jobs will fundamentally be altered by AI application where the cognitive literacy of every related stakeholder will need to be improved (Sharma & Rana, 2021; Holtel, 2015).

Personality assessment questionnaires can also help organizations to determine which current employees are suited for new positions as the organization changes.

Impact of AI on Global Business Practices

An organization looking to fill cyber-security positions, for example, may already have employees with the skills needed to fill those positions. Internal hiring would be more efficient and cost-effective than sourcing candidates from outside sources.

Walmart is among one of the top companies which has applied AI in its all areas of operation where it does automated inspections, catching items that humans may miss and freeing them to do other things. Netflix, for example, uses customization and personalization to curate content for specific customers based on their previous viewing habits. HR managers may use the same principles to create an adaptive learning experience with a similar feel and look, curating content for training from both external and internal sources based on preferences, career goals, personality insights and work history.

AI in marketing and sales: Artificial intelligence has progressed from science fiction to commercial reality. You need AI solutions that integrate with your infrastructure and data strategy to meet today's challenges and prepare for the future. Amazon's Alexa and IBM's Watson are ideal examples of the existence of artificial intelligence which is already creating its way in marketing. Google assistance will help fulfill various needs like buying a present, booking a hotel, searching for a good college, etc. IBM's Watson computer has the efficiency to read 200 million pages of data within a second and manage to predict the buyer's tendency in the market (Howard, 2014). To redirect the advertising and create recommendations, Facebook takes the information from web browsers. LinkedIn is another example of using machine learning to find talent and align workers and employers, as it incorporates the accounts of everyone in the user's network, as well as their interests. An organization can be competent with others with the help of AI which provides market information, as marketing automation helps a company to expand its value and impacts its content, improves leads-to-sale conversation rates, captures lead intelligence, drives repeat purchasing and, most significantly, boosts the overall customer experience throughout its journey in the market (Roetzer, 2014).

The chances of AI taking over marketing completely are very high, because marketing programs not only need to know their customers, but also need to know the program their customers are using to make their decisions; and with the use of automation, AI advancements may also help marketers to interact with humans and capture their buying decisions.

Customer retention and sales forecasting are some of the functions of artificial intelligence which are already in use. Some more functions can also be run smoothly with the involvement of artificial intelligence in marketing and sales like sales transactions, regional sales distribution, the involvement of sales executive customer incident data and mining of customer actions and social sentiments—all can be used to forecast sales and provide the positive influence to a company, helping in determining customers who are at high risk of leaving (Computer, 2017).

AI in production and job markets: Adaptation and innovation are the two most important factors in the manufacturing sector, where development should lead to sustainable production using the latest technologies. For the promotion of sustainable production, there is a need for smart production from a global perspective with smart technologies. In this direction, artificial intelligence has already been established in the manufacturing section via machine learning and automation to achieve sustainability in manufacturing (Cioffi et al., 2020). Nevertheless, AI is likely to increase the transition toward service economies, and will also have an impact on the type and the quality of economic growth—but the outcome of the impact of AI and jobs is still doubtful, especially in developing countries, where contrarily it is likely to expand and automation and speed up job losses for low-skilled, blue-collar workers in manufacturing fields (Arnet et al., 2016); nonetheless, it will compel specific worker skills to be critical for adding value to production and goods, resulting in a further expansion of the share of production and services not only at a national level but also at an international level, resulting in more job opportunities. The Xeon processor, Xeon Phi processor, Xeon Phi E5 processor, Stratix FPGA, Arria FPGA, Max FPGA, Cyclone FPGA, Enpirion FPGA, Neon platform, Nervana Cloud, Nervana Engine, and Saffron platform are among the AI hardware and platforms developed by Intel Corporation.

This is the optimistic perspective which suggests the possibility of the creation of more jobs, but for skilled and trained people. Financial services firms all over the world are embracing artificial intelligence and intelligent process automation. Customer expectations are increasing, and financial services firms can achieve a competitive advantage by automating customer service. Thus, there is no doubt that AI has opened a new door for more job opportunities, but there is a need for workers to change or develop their skills to be more successful in the emerging global economy—and there will be an employment shift.

As per one of the studies published by Oxford University, 47% of jobs in the United States are inclined toward automation, where the majority of these jobs to automation is from white-collar people such as auditors, officers, financial analysts, insurance companies and accountants; on the other side, the most difficult-to-replace jobs are therapists, chief executive officers from the blue-collar zone. The most vulnerable ones are high-labor jobs such as drivers, gamers and cashiers, and least vulnerable are computer scientists, writers, editors, designers and lawyers (Kaplan, 2016). Google is one of the most well-known companies when it comes to artificial intelligence. Despite the fact that Google does not store the information of millions of people in its search engine, it does use it to redirect your subsequent searches. As soon as you start writing a few letters, it guesses what you are looking for and uses predictive text to make suggestions.

6.4 INFERENCES

At present, artificial intelligence is at the topmost list in terms of the technology requirement all over the world, where there is still a disturbance with workers at the present work which is disturbed by innovation. Besides some drawbacks, AI has more benefits, such as the following.

1. **More transparency at the workplace:** AI will be useful in the identification of errors and issues of deception in the auditing process. Fear of failure is the most common issue with industry people. With the help of AI, the chances of errors and omission will be less and the chances of success by maintaining the standards of ethics and transparency will be greater.

Impact of AI on Global Business Practices 109

Automation of data will provide greater control to determine the weak links that are responsible for poor quality standards and workers' welfare. This transparency will not only give connectivity of workers to their organization, but also to customers, who can get product information, better customer service and quick responses. People connected with the organization will be able to access core data, as well, and will have more connectivity. More clarity at the workplace will create more belongingness and association, and finally, a healthy work environment leads to more productivity, profit, expansion and job opportunities (Sharma et al., 2021a).

2. **More chances of investment returns:** More transparency and productivity will assure competitive advantage and increase efficiency and productivity. Use of AI technology will build faster and cheaper production process in business, and an organization could have the first-mover advantage at international level. Here, there is a guaranteed investment return.

3. **Fewer chances of human errors:** Use of automation in business results in less human error and improvement in efficiency and cost-effectiveness, and this will add to the continuity of production without any obstacles. A well-established technology could help minimize waste and maximize quality by reducing the chances of human error.

4. **More teamwork and collaborations:** Artificial intelligence can communicate more with employees in a better manner and will improve the quality of the workforce. Integration of the workforce makes the job simpler and easier if supported by technology. Overburden of work will be reduced, and more transparency can help in smooth work of teams and collaborations.

5. **Better human wellbeing:** AI technologies and innovations will also start affecting workers' quality of life directly with fewer working hours and greater payments. When skills and training will be compulsory, the standard of living of workers and their wellbeing will improve. More profits, more expansion and more job opportunities will give more options to workers to make their lives much better.

6. **More job opportunities at the international level:** When we compare developed nations with the developing ones, developed nations are always having scarcity of workers while developing nations have a surplus. AI and ML will be the milestones to change the global business scenario. The global business world will perform in a much better way with the involvement of AI and automation.

6.5 CONCLUSION

Artificial intelligence has always been taken as hype, especially in developing nations, but it has the competency to transform the global economy with technological innovations and scientific knowledge in entrepreneurial activities. AI technologies are liable to lead to great automation and more connectivity; hence, AI is taking this world into the Fourth Industrial Revolution, which will have impact on individuals, companies, communities, governments and, of course, global business. Different people have different perspectives on AI. At present, AI is used for information and automation, but it has fewer learning capabilities, although it will take so many years to accept its full heart as it is still a threat to the unskilled workforce at the initial level and all humanity at the next level. Such stage would come after reaching the peak, as AI has potential to rebuild skill demand, more career opportunities and better distribution of workers among industries. Thus, emerging economies should focus on skill development to achieve a competitive advantage in global business.

REFERENCES

Agarwal, S. (2018). *Block Chain Technology in Supply Chain and Logistics*. Cambridge, MA: Massachusetts Institute of Technology.

Andrews, S., Ayers, S., Bakovic, T., et al. (2019). Reinventing Business Through Disruptive Technologies: Sector Trends and Investment Opportunities for Firms in Emerging Markets. *IFC Report*. Retrieved from www.ifc.org/wps/wcm/connect/8c67719a-2816-4694-91877de2ef5075bc/Reinventing business-through-Disruptive-Tech- v2.pdf?MOD= AJPERES&CVID=mLo6cfr

Armstrong, S., Sotala, K., & Ó hÉigeartaigh, S.S. (2014). The Errors, Insights and Lessons of Famous AI Predictions – and What they Mean for the Future. *Journal of Experimental & Theoretical Artificial Intelligence*, 26(3), 317–342.

Arnet, M., Gregory, T., & Zierahn, U. (2016). The Risk of Automation for Jobs in OECD Countries: A Comparative Analysis. *OECD Social, Employment and Migration Working Papers No. 189.*

Arora, B., & Rahman, Z. (2017). Information Technology Capability as Competitive Advantage in Emerging Markets: Evidence from India. *International Journal of Emerging Markets*, 12(4), 447–463.

Atzori, L., Iera, A., & Morabito, G. (2010). The Internet of Things: A Survey. *Computer Networks*, 54(15), 2787–2805. http://doi.org/10.1016/j.comnet.2010.05.010

Belvedere, V., Grando, A., & Bielli, P. (2013). A Quantitative Investigation of the Role of Information and Communication Technologies in the Implementation of a Product-service System. *International Journal of Production Research*, 51(2), 410–426. http://doi.org/10.1080/00207543.2011.648278

Bloem, J., van Doorn, M., Duivestein, S., Excoffier, D., Maas, R., & van Ommeren, E. (2014). *The Fourth Industrial Revolution: Things to Tighten the Link Between It and OT*. VINT Research Report 3 of 4. 2014 Groningen: Sogeti VINT Production LINE UP Boeken Media BV.

Brynjolfsson, E., et al. (2017). *Artificial Intelligence and the Modern Productivity Paradox: A Clash of Expectations and Statistics*. NBER Working Paper No. 24001, p. 10.

Bughin, J., Hazan, E., Ramaswamy, S., Chui, M., Allas, T., Dahlström, P., Henke, N., & Trench, M. (2017). *Artificial Intelligence the Next Digital Frontier?* McKinsey and Company Global Institute, Discussion Paper.

Cao, L. (2017). Data Science: A Comprehensive Overview. *ACM Computing Surveys*, 50(3), 1–42. https://doi.org/10.1145/307625

Cioffi, R., Travaglioni, M., Piscitelli, G., Petrillo, A., & De Felice, F. (2020). *Department of Engineering*, Parthenope University, Isola C4, Centro Direzionale, 80143, Napoli NA, 8 January 2020.

Coats, P.K. (1987). Artificial Intelligence, Expert Systems and Business. *American Business Review*, 5(2), 7p.

Impact of AI on Global Business Practices 111

Computer, E. (2017). *Artificial Intelligence Transforming the Future of Work*. Express Computer. Retrieved from https://searchproquestcom.nnu.idm.oclc.org/docview/1911103619

Davenport, T., & Ronaki, R. (2018). Artificial Intelligence for the Real World. *Harvard Business Review*, 4–10.

Deng, L., & Yu, D. (2014). Deep Learning: Methods and Applications. *Foundations and Trends® in Signal Processing*, 7(3–4), 197–387. http://doi.org/10.1561/2000000039

Frank, R., Autor, D., Bessen, J.E., Brynjolfsson, E., Cebrián, M., Deming, D., Feldman, M., Groh, M., Lobo, J., Moro, E., Wangk, D., Younk, H., & Rahwan, I. (2019). *Toward Understanding the Impact of Artificial Intelligence on Labor*. Edited by Jose A. Scheinkman. New York: Columbia University.

Fredman, J. (2018). The Role of Accounting & Finance in Business Management. *Chron*. Retrieved from http://smallbusiness.chron.com/role-accounting-finance-business-management65620

Goodfellow, I., Bengio, Y., & Courville, A. (2016). *Deep Learning*. Cambridge, MA: MIT Press.

Haider, P., Chiarandini, L., & Brefeld, U. (2012). Discriminative Clustering for Market Segmentation. Proceedings of the *18th ACM SIGKDD International Conference on Knowledge Discovery and Data Mining*. ACM.

He, K.F. (2016). *Cyber-Physical System for Maintenance in Industry 4.0*. Master's Thesis. Jonkoping University.

Holtel, S. (2015). Artificial Intelligence Creates a Wicked Problem for the Enterprise. *Procedia Computer Science*, 99, 171–180.

Howard, J. (2014). *The Wonderful and Terrifying Implications of Computers that Can Learn*. Retrieved from www.ted.com/talks/jeremy_howard_the_wonderful_and_terrifying_implications_of_computers _that_can_learn

Kagermann, H., Wahlster, W., & Johannes, H. (2013). *Recommendations for Implementing the Strategic Initiative INDUSTRIE 4.0*. Forschungsunion.

Kaplan, J. (2016). *Artificial Intelligence: What Everyone Needs to Know*. Oxford: Oxford University Press.

Klosters, D. (2016). World Economic Forum Annual Meeting 2016 Mastering the Fourth Industrial Revolution. *World Economic Forum*. Retrieved from http://www3.weforum.org/docs/Media/

Liao, Y., Loures, E. R., Deschamps, F., Brezinski, G., & Venâncio, A. (2017). The Impact of the Fourth Industrial Revolution: A Cross-country/Region Comparison. *Production*, 28, e20180061. https://doi.org/10.1590/0103-6513.20180061

Maynard, A. D. (2015). Navigating the Fourth Industrial Revolution. *Nature Nanotechnology*, 10(12), 1005–1006. https://doi.org/10.1038/nnano.2015.286. Retrieved from www.ncbi.nlm.nih.gov/pubmed/26632281

McAfee, A. (2013). *What Will Future Jobs Look Like*? Retrieved from www.ted.com/talks/andrew_mcafee_what_will_future_jobs_look_like#t- 166952

Mosterman, P., & Zander, J. (2015). Industry 4.0 as a Cyber-Physical System Study. *Software & Systems Modeling*, 15, 17–29.

Mou, Xiaomin. (2019). Senior Investment Officer, Private Equity Funds—Disruptive Technologies and Funds. IFC, Washington, DC.

National Academy of Science and Engineering. (2013). *Acatech Annual Report, 2013*. Retrieved from https://en.acatech.de/publication/jahresbericht-2013

Nilsson, N. (2010). *The Quest for Artificial Intelligence: A History of Ideas and Achievements*. Cambridge: Cambridge University Press.

Nisbet, R., Miner, G., & Yale, K. (2018). Deep Learning, Chapter-19. In *Handbook of Statistical Analysis and Data Mining Applications* (2nd ed.). Cambridge, MA: Academic Press.

Park, S. C. (2017). The Fourth Industrial Revolution and Implications for Innovative Cluster Policies. *AI & Society*, 33(3), 433–445.

Parkes, D. C., & Wellman, M. P. (2015). Economic Reasoning and Artificial Intelligence. *Science*, 349(6245), 267–272.

Pavaloiu, A. (2016). *The Impact of A.I. of Global Trends*. Retrieved from www.biblio.psbedu.net

Prakash, C., Saini, R., & Sharma, R. (2021). Role of Internet of Things (IoT) in Sustaining Disruptive Businesses. In R. Sharma, R. Saini, & C. Prakash, V. Prashad (Eds.), *Role of Internet of Things (IoT) in Sustaining Disruptive Businesses* (1st ed.). New York: Nova Science Publishers.

Rana, G. (2010). Knowledge Management and E-Learning Activities in the 21st Century to attain Competitive Advantage. *Advances in Management*, 3(5), 54–56.

Rana, G., & Sharma, R. (2019). Emerging Human Resource Management Practices in Industry 4.0. *Strategic HR Review*, 18(4), 176–181.

Remes, J., Manyika, J., Bughin, J., Woetzel, J., Mischke, J., & Krishnan, M. (2018). *Solving the Productivity Puzzle: The Role of Demand and the Promise of Digitization*. McKinsey Global Institute.

Rodič, B. (2017). Industry 4.0 and the New Simulation Modelling Paradigm, *Organizacija*, 50(3), 193–207.

Roetzer, P. (2014). *The Marketing Performance Code: Strategies and Technologies to Build and Measure Business Success*. Hoboken, NJ: John Wiley & Sons, Inc.

Sadek, A. W., & Chowdhury, M. (2012). *Artificial Intelligence Applications to Critical Transportation Issues*. Retrieved from www.researchgate.net/profile/Said_Easa/publication/273576102_Design_and_construction_of_transportation_infrastructure_httponlinepubstrborgonlinepubscircularsec168pdf/links/55097a910cf26ff55f85932b.pdf#page=14

Saturno, M., Pertel, V. M., Deschamps, F., & Loures, E. R. (2017). Proposal of an Automation Solutions Architecture for Industry 4.0. In Proceedings of the *24th International Conference on ge Production Research*. Poznan: ICPR.

Schwab, K. (2017). *The Fourth Industrial Revolution*. Geneva: World Economic Forum.

Sharma, R., & Rana, G. (2021). Revitalizing Talent Management Practices through Technology Integration in industry 4.0. In R. Sharma, R. Saini, & C. Prakash (Eds.), *Role of Internet of Things (IoT) in Sustaining Disruptive Businesses* (1st ed.). New York: Nova Science Publishers.

Sharma, R., Rana, G., & Agarwal, S. (2021a). Techno Innovative Tools for Employer Branding in Industry 4.0. In: G. Rana, S. Agarwal, & R. Sharma (Eds.), *Employer Branding for Competitive Advantage* (1st ed.). Boca Raton, FL: CRC Press. https://doi.org/10.4324/9781003127826-11.

Sharma, R., Saini, A. K., & Rana, G. (2021b). Big Data Analytics and Businesses in Industry 4.0. *Design Engineering*, 2021(2), 238–252.

Sırmaçek, B. (2007). *FPGA İle Mobil Robot İçin Öğrenme Algoritması Modellenmesi*. İstanbul: Yükseklisanstezi, Yıldız TeknikÜniversitesi.

Smith, C., McGuire, B., Huang, T., & Yang, G. (2017). *The History of Artificial Intelligence* [Scholarly Project]. Retrieved November 20, 2017 from https://courses.cs.washington.edu/courses/csep590/06au/projects/history-ai.pdf

Soni, Neha, Sharma, E., Singh, N., & Kapoor, A. (2019). Impact of Artificial Intelligence on Businesses: From Research, Innovation, Market Deployment to Future Shifts in Business Models. Final manuscript is submitted to *Journal of Business Research—Elsevier for Consideration*.

Impact of AI on Global Business Practices

Strusani, D., & Houngbonon, G. V. (2019). The Role of Artificial Intelligence in Supporting Development in Emerging Markets. *EMCompass*, No. 69, International Finance Corporation, Washington, DC.

Tantuğ, A. C., & Türkmenoğlu, C. (2015). *Türkçe Metinlerde Duygu Analizi*. İstanbul: Yükseklisanstezi, İstanbul TeknikÜniversitesi.

Turing, A. (1950). Computing Machinery and Intelligence. *Mind*, 59, 433–460.

Van Gordon, W., Shonin, E., Zangeneh, M., & Griffiths, M. D. (2014). Work-Related Mental Health and Job Performance: Can Mindfulness Help? *International Journal of Mental Health and Addiction*, 12, 129–137.

Wasserman, P. (2016). *Santander InnoVentures Distributed Ledger Challenge: Decoding Block Chain*. Retrieved from www.sachsinsights.com/santanders-innoventures-distributed-ledger-challenge-decoding- blockchain

Witkowski, K. (2017). Internet of Things, Big Data, Industry 4.0—Innovative Solutions in Logistics and Supply Chains Management. *Procedia Engineering*, 182(1), 763–769.

Xu, X. (2012). Robotics and Computer-Integrated Manufacturing From Cloud Computing to Cloud Manufacturing Ubiquitous Product Life Cycle Support. *Robotics and Computer Integrated Manufacturing*, 28(1), 75–86.

Zhong, R. Y., Xu, X., Klotz, E., & Newman, S. T. (2017). Intelligent Manufacturing in the Context of Industry 4.0: A Review. *Engineering*, 3(5), 616–630.

7 Road Map for Implementation of IoT in Metal Cutting for the Process Monitoring System to Improve Productivity

Mukhtar Sama and Ashwini Kumar Saini

CONTENTS

7.1 Introduction 116
    7.1.1 Dynamic and Self-Adapting 117
    7.1.2 Self-Configuring 118
    7.1.3 Interoperable Communication Protocols 118
    7.1.4 Unique Identity 118
    7.1.5 Integrated Into the Information Network 118
7.2 Literature Review 118
7.3 Challenges of Implementing IoT in Metal Cutting 120
7.4 Stages of IoT in Metal Cutting 120
    7.4.1 Problem Definition 121
    7.4.2 Data Collection 121
        7.4.2.1 Automatic 121
        7.4.2.2 Semi-Automatic 121
        7.4.2.3 Sensor and Communication Technology 121
        7.4.2.4 Sensors for Metal Cutting 122
        7.4.2.5 Gateway and Protocols 122
    7.4.3 Data Preprocessing 122
        7.4.3.1 Data Preparation 124
        7.4.3.2 Data Reduction 124
    7.4.4 Model 124
        7.4.4.1 Descriptive Analytics 124
        7.4.4.2 Diagnostic Analytics 124
        7.4.4.3 Predictive Analytics 125
        7.4.4.4 Perspective Analytics 125

DOI: 10.1201/9781003145011-7

7.4.5	Decision-Making	126
7.4.6	Implementation	126
7.4.7	Monitoring and Feedback	126
7.5	Case Study—Lathe Machine	126
7.6	Measuring the Value-Added Time	127
7.7	Measuring the Non–Value-Added Time	127
7.7.1	Setup Time	127
7.7.2	Inevitable Stoppages	128
7.7.3	Maintenance	128
	7.7.3.1 Preventive Maintenance	128
	7.7.3.2 Planned Maintenance	128
	7.7.3.3 Predictive Maintenance	129
	7.7.3.4 Autonomous Maintenance	129
	7.7.3.5 Breakdown Maintenance	129
7.7.4	Obligatory Stoppages	129
7.8	Sensors for Quality Improvement and Conditioning Monitoring	130
7.9	Data Application	130
7.9.1	Production Department	130
7.9.2	Planning Department	131
7.9.3	Quality Management	131
7.9.4	Maintenance Management	132
7.10	Conclusion	132
7.11	Limitations of the Study	132
7.12	Future Scope	133
References		133

7.1 INTRODUCTION

Since the First Industrial Revolution, machining has been the most critical manufacturing process. Metal cutting is needed in almost every sector, like household items, medical, automotive, defense, aerospace, etc. Primary processes like casting and forming fail to give higher dimensional accuracy and surface finish. The main requirement to carry out metal cutting operations is to achieve accuracy and surface finish. It is also used to achieve the final shape of the components. Primitive humans used a cutting tool made of stone and bone to cut the various materials. Egyptian civilization used the machining operation to drill holes in the rock of pyramids. The use of machining processes is also found in Indian civilization; lathe machines were used for carving the temple pillars. In South India, entire parts of the temple are made by machining a single stone. Before the First Industrial Revolution, most machine tools for machining were run by flowing water, wind or animal energy (Frader, 2005). Due to the limitation of the power, it was not possible to cut hard materials like carbon steel and cast iron. The invention of the steam engine boosted the machine tool industries due to the availability of high power. The First Industrial Revolution took place due to the invention of the steam engine. Steam engines required ample space, and they also needed a boiler. Starting time of the steam engine is also significant. To overcome these limitations, an internal combustion engine that is portable and easy to start was used in place of a steam engine in the First Industrial Revolution; due to the use of the manual machine in the first industrial revolution the lower production rate and lower quality were the biggest challenges. To overcome the production rate and poor quality of machined components, cam and follower–based hard automation machines like capstans and turret lathes were introduced in the Second Industrial Revolution. Due to the more compact and efficient electric motor in the nineteenth century, the electric motor started as the prime mover of the machine tool. However, due to the machine tool's hard automation, it was challenging to adapt product design changes. Soft automation, which can be achieved by computer numerical control (CNC) machine tools and programmable logic control (PLC) systems, was introduced to reduce the changeover time from one product to another product in the Third Industrial Revolution, which began after the invention of the computer in 1960 (Cooper and Kaplinsky, 2005). The remote control and monitoring of CNC machine tools were not possible; the set of instructions in the form of code need to be given by an operator for various operations.

In 2012, internet technology was introduced by Germany to control and monitor the process remotely. Also, the various data of the operations can be collected using a sensor, and that data can be stored in a central server. This stored data can be used for decision-making and improving the efficiency of processes. The implementation of internet technology in manufacturing is known as the Fourth Industrial Revolution or Industry 4.0. Industry 4.0 is about using internet technology for remote control, monitoring and automated decision-making using the collected data. In the mass production industries, activities are fixed while in job shop production system the activities are not fixed (Alptekinoğlu and Corbett, 2008). It is also a challenge to carry out the time study in the project-based industry due to different sets of activities (Vinod and Sridharan, 2008). India, an emerging economy, is proving to be a vast opportunity market for the world's big manufacturing giants (Trivedi, 2015). Limited research has been done on the time study of metal cutting using the Internet of Things (IoT). In this study, the integration of humans and IoT is discussed to monitor the time of various activities on the metal cutting machine. The stages of implementing the IoT with its necessities and challenges in metal cutting are discussed. An overview of IoT technologies from the aspect of data acquisition, data preprocessing and data analytics is also discussed.

IoT is all about connecting the devices and servers to store the data. The IoT device must have the following characteristics.

1. Dynamic and self-adapting
2. Self-configuring
3. Interoperable communication protocols
4. Unique identity
5. Integrated into the information network

7.1.1 Dynamic and Self-Adapting

The IoT device must be capable of adapting to the change in condition. Mobile display light changes according to the amount of light in the room are the best example of dynamic and self-adapting conditions.

7.1.2 Self-Configuring

In the IoT system, there are large numbers of sensors and actuators. To provide specific functionality, sensors are interconnected, so the IoT devices must have self-configuring capabilities.

7.1.3 Interoperable Communication Protocols

It may be possible that in the IoT system, all devices are not operating on the same communication protocols. IoT devices must be designed in such a way that they can work in multiple communications protocol environments.

7.1.4 Unique Identity

To send the data to an exact device must have a unique identity. It is nothing but the residential addresses of the individual.

7.1.5 Integrated Into the Information Network

All the sensors and radio frequency identification (RFID) data must be integrated into the information network that allows them to communicate and exchange the data with each other and systems

The use of internet technology in domestic products began many years ago, like watches, air conditioning, refrigerators, etc. According to the U.S. National Intelligence Council, by 2025 everything in our life will be connected via the internet. The use of IoT in manufacturing can improve the efficiency of the industry. But there are many challenges to implement IoT in manufacturing. Dai and colleagues studied the challenges like harsh environment, high data rate, delay-sensitivity, and many more (Dai et al., 2020).

7.2 LITERATURE REVIEW

Big Data analytics is an essential tool in the Fourth Industrial Revolution. Big Data analytics has three stages: data acquisition, data preprocessing and storage, and data analytics (Dai et al., 2020; Sharma et al., 2021a). Data analytics has four applications: descriptive analytics, diagnostic analytics, predictive analytics and prescriptive analytics. Descriptive analytics gives us the status of the system. Diagnostic analytics is used to find out the root cause of the problem, like the machine tool's failure. To predict the runtime of the machine tool, we can use predictive analytics. To achieve the target production, we can use perspective analytics. Data collection is done by sensor and communication technology. Competitive advantage in smart manufacturing can be taken over a traditional manufacturing system by the planned investment. The ten performance dimensions like cost, quality, flexibility, time, integration optimized productivity, real-time diagnosis and prognosis, computing, social and ecological sustainability are considered to develop a smart manufacturing performance measurement system to

Road Map for IoT in Metal Cutting

evaluate the smart manufacturing system (Kamble et al., 2020). Alvian and colleagues introduced the concept of a programmable manufacturing advisor (PMA) to automate the decision-making in smart manufacturing. PMA consists of three units: information units, analytical unit, and optimization units. Information units continuously update the parameter (Alvian et al., 2020). It is challenging to decide the priority of maintenance activity in large-scale industries.

Direct and indirect sensing techniques are mainly used in metal cutting operations. Vision-based sensing is an example of a direct sensing technique, and acoustic emission is an indirect technique (Liang et al., 2004). An IoT-based industrial data management system (IDMS)—which consists of five layers: physical, network, middleware, database and application layer—can effectively collect and analyze the raw industrial data and urgent events by using the communication protocols (Saqlain et al., 2019). Digitalization of manufacturing industries can reduce the product cost and flexibility of the production system by integrating data and value chain in product design, production processes and resources (Perzylo et al., 2020). Using real-time sensor data and simulation, a digital twin can be developed which can be used to make a digital replica of physical devices for monitoring purposes.

Due to globalization, manufacturing companies cannot increase the selling cost of their products. Companies try to increase profit by increasing the efficiency of their systems. Efficiency can be increased by minimizing the waste (Sharma et al., 2021b). In lean manufacturing, there are many tools and techniques available to reduce waste in industries—one of the ways to identify waste in the system is value stream mapping (VSM). It gives us a bird's-eye view of the system, with which we can easily differentiate between the value-added and non–value-added activities (Venkataraman et al., 2014). VSM is a graphical representation of all the activities of the system. Generally, various activities are recorded manually, and then that data is used to make the VSM. In metal cutting, measuring the time of various activities is challenging. In the metal cutting operation, the time during which cutting occurs is considered value-added, and all other activities are considered non–value-added activities.

Next-generation intelligent controllers are capable of real-time monitoring, diagnosis, self-learning and adaptive optimization (Hentz et al., 2013). Smart machining utilizes a machine learning algorithm that can be trained using the machining data to improve the product quality, increase productivity and monitor the system's health, and many more (Kim et al., 2018). The smart cutting tool can be designed using piezoelectric film to measure the cutting force during the machining process. The adaptive control system can be developed by integrating the smart tool with a CNC machine (Wang et al., 2013). The intelligent algorithm can suggest the optimum feed rate to maximize the material removal rate (MRR) in the smart machining (Park et al., 2018). The integration of finite element method (FEM), high-speed light and thermal camera, and image processing can give us more insight into the machining processes (Deshayes et al., 2006). A self-optimizing control system that predicts feed using an artificial neural network (ANN) can be an efficient tool for improving product quality. The input parameters used to train the ANN are tool wear, cutting speed and depth of cut (Park and Tran, 2014). To detect the on–off condition of machine tool, the vibration and Gaussian mixture model can be used (Wang et al., 2020).

The cutting force model can decide the safer and faster-cutting parameters of machining during the process planning (Jerard et al., 2009).

7.3 CHALLENGES OF IMPLEMENTING IOT IN METAL CUTTING

The IoT objectives in metal cutting are to improve productivity, quality, capability, overall equipment efficiency (OEE), remote monitoring and control of the machine tool. The harsh environment due to high vibration and dust is the biggest challenge in implementing IoT in metal cutting industries. Volume, velocity, and variety are three major phalanges regarding the data collection in Industry 4.0. The problem of a variety of data can be solved by (Zhang et al., 2016). Also, there are many types of critical parameters—like cutting speed, feed, depth of cut, temperature, and many more—need to be stored to improve the machine tool's efficiency in the metal cutting operation. The highly dynamic nature of the metal cutting process generates data at a very high rate. So the velocity of data in metal cutting is also a big challenge. The volume of data generated by multiple machine tools is too large.

7.4 STAGES OF IOT IN METAL CUTTING

It is not easy to convert existing traditional industries into IoT-enabled industries. It becomes more complex in the job shop production industries, where the product design and demand are not fixed. The following road map (Figure 7.1) can be used to implement IoT in metal cutting industries.

1. Problem definition
2. Data collection
3. Data preprocessing
4. Modeling
5. Decision-making
6. Implementation
7. Monitoring and feedback

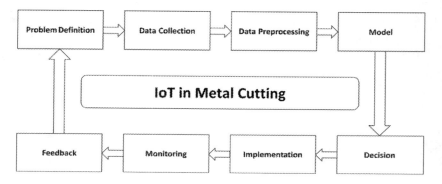

FIGURE 7.1 Road map for implementing IoT in metal cutting.

Road Map for IoT in Metal Cutting

7.4.1 PROBLEM DEFINITION

The objective of using IoT in industries differs from industry to industry. Lower production rate, poor quality, on-time delivery, downtime of machine tools, and many more are the most significant issues of project-based industry, where a wide variety of products exist. It is vital to identify the list of problems industries want to be addressed with the help of IoT. The list of the problem must be shortlisted.

7.4.2 DATA COLLECTION

The main objective of the application of IoT in machining is for automation, process monitoring and control, and predictive maintenance (Benardos and Vosniakos, 2017). It is essential to shortlist tentative responsible parameters to control the variety of data. Filtration of the critical parameters takes place in the preprocessing stages by using various data preprocessing algorithms. After identifying the required parameter, the next step is how to collect the data. The data collection can take place in the following two different ways.

7.4.2.1 Automatic

The majority of the parameters collection should be done in automatic for high accuracy.

The temperature, vibration, speed, etc., can be collected automatically with the help of sensors. RFID can be used to identify the machine tool or product automatically.

7.4.2.2 Semi-Automatic

Some of the parameters cannot be collected solely from sensors, mainly in the job shop production system. Such parameters can be stored using the human–machine interface (HMI). In HMI, the operator needs to provide input. Due to human intervention, data is less accurate.

Collected data can be classified into three categories: 1) structured data, 2) unstructured data, and 3) semi-unstructured data. Sensor and RFID data in Microsoft Excel form are an example of structured data, while video, audio and images are examples of unstructured data. XML and JSON files are examples of semi-structured data. The biggest challenge to send the data to the server is high volume, high verity and high velocity.

7.4.2.3 Sensor and Communication Technology

The sensor is the bridge between physical information and digital information. The use of sensor networks has already been implemented in sectors like health, supply chain, defense, etc. (Atzoria et al., 2013). Wireless sensor networks (WSNs) work in low bandwidth and delay tolerant applications ranging from construction and defense to environmental and healthcare monitoring. WSN nodes consist of a low-power sensing device, embedded processor, communication channel and power module. The embedded system collects and processes the input signal. Change in the physical condition like temperature or speed is measured and converted into a sensor's digital signal. Real-time movement of components on the shop floor can be traced with RFID technology. Integrating different technologies of RFID in the production system is the biggest challenge in the industry (Zhang et al., 2011).

7.4.2.4 Sensors for Metal Cutting

Lee and colleagues discuss the sensor's use in the machining process like encoder force meters, laser interferometers, accelerators and acoustic emissions in metal cutting (Lee and Dornfeld, 1998). A study of critical elements of equipment and its parameter was carried out by Urbikain and Lacalle (2020) which identifies essential elements: spindle head, lubrication, linear axis, and many more. Power consumption is the main critical cost factor in manufacturing processes, so it is vital to monitor the power consumption of manufacturing processes. IoT-based energy monitoring of the machining process can be done using HC33C3 power sensors (Chen et al., 2018). To predict the flank wear in the drilling process, sensors like dynamometers, current sensors and accelerometers can be used (Patra et al., 2013). Vibration during the machining causes chatter which leads to poor surface finish. Khalifa and colleagues discussed the monitoring of chatter through image processing (Khalifa et al., 2006). A complementary metal oxide semiconductors (CMOS) can acquire frames faster, but the sensitivity is lower than when using a charge-coupled device (CCD) (Dutta et al., 2013).

7.4.2.5 Gateway and Protocols

Gateway is the filtering device between devices and the cloud. It also does primary preprocessing at the local level to optimize the data before sending it to the server. Another application of gateway is as a security check to make sure that only genuine data goes to a server or comes from the server. Digital data is collected from various types of sensors. IoT demands ubiquitous connectivity to integrate multiple heterogeneous networks, such as Zigbee ad hoc networks, wireless LAN, cable networks, etc. Sectors like automotive, healthcare, transportation, environmental monitoring, etc., needs smart gateways to facilitate high data rates, end-to-end connectivity and higher bandwidth of multi-hop networks among heterogeneous networks when the devices require communicating with each other and server (Guoqiang et al., 2013). For data exchange, the device must for the set of rule which is known as communication protocols.

The following are some examples of protocols: 1) 802.3-Ethernet, 2) 802.11-WiFi, 3) internet protocol version 4(IPv4), 4) internet protocol version 6(IPv6), 5) transmission control protocol (TCP), and 6) hypertext transfer protocol (HTTP).

7.4.3 Data Preprocessing

The raw data collected in the server cannot be used directly due to missing data, wrong data, redundant data, etc. To get useful information from the stored data, some data preprocessing is required (Figure 7.2). There are mainly two objectives in data preprocessing: 1) data preparation, and 2) data reduction.

Road Map for IoT in Metal Cutting

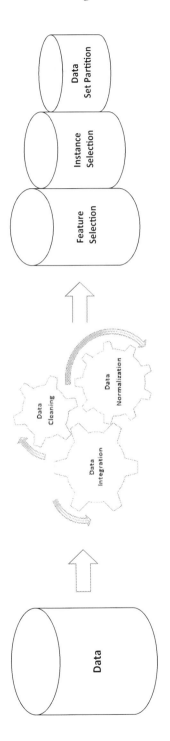

FIGURE 7.2 Data preprocessing techniques.

7.4.3.1 Data Preparation

Data preparation tasks are mandatory to perform; otherwise, the information provided by these data may lead us to a wrong decision. Data cleaning tasks involve the detection of discrepancies and dirty data. Data transformation combines the original raw attributes using different mathematical formulas originated in the manufacturing system or pure mathematical formulas. Data is collected through various sensors in a different form from multiple machines in the industry, so this data needs to be integrated by the data integration method. Collected data from the sensor have different scales; for example, the RPM (revolutions per minute) data is enormous, while data of the depth of cut is tiny. Generally, data is normalized in the range of 0 to 1 and −1 to 1.

7.4.3.2 Data Reduction

Since the data collected is vast and there are many attributes contained, to improve the computation accuracy and to reduce the computational effort, it is very important to identify the important attributes and set of instances.

7.4.4 MODEL

Collected data can be analyzed using various statistical and machine learning algorithms (Figure 7.3). Collected data can be used for the following four purposes.

1. Descriptive analytics
2. Diagnostic analytics
3. Predictive analytics
4. Perspective analytics

7.4.4.1 Descriptive Analytics

In IoT-based industries, hundreds of sensors send the data. It is essential to infer meaningful information. Descriptive analytics is the lowest level of analytics. It is a tool for preparing a summary from the past data which can be easily understood. Mean, median, mode and the standard deviation are the tools of descriptive analysis. Pattern recognition is also an important part of descriptive analytics.

7.4.4.2 Diagnostic Analytics

Diagnostic analysis gives more detail as compared to descriptive analytics. Diagnostic analytics helps determine the root cause of the problems we have identified by descriptive analytics. As compared to descriptive analytics, diagnostic analytics requires high computation effort. Adaptive neuro-fuzzy interface system (ANFIS) combines the advantage of neural networks and fuzzy systems. ANFIS is a very efficient tool for fault detection of multi-component systems (Moeuf et al., 2018). Stack denoising autoencoder (SDAE) is an effective and reliable deep learning tool for fault detection (Yu et al., 2019). The decision tree and clustering techniques are also used to detect faulty conditions (Kozjek et al., 2017).

Road Map for IoT in Metal Cutting

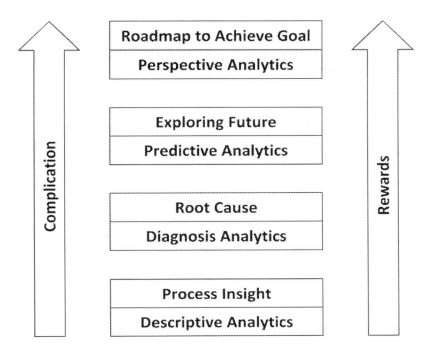

FIGURE 7.3 Complication vs. rewards.

7.4.4.3 Predictive Analytics

Predictive analytics is used to forecast current analytics. This analytics is computationally more complex as compared to the descriptive and diagnostic techniques. Regression modeling is the most common method use for predictive analytics. The mathematical relationship is established between input and output using past data. Regression analysis fails to give accurate predictions where a complex relationship exists between parameters. The artificial neural network can handle the complex relationship which is developed based on a biological neural network. K-nearest neighbor (KNN) is the method to predict the class of the data. Grolinger and colleagues compare the support vector machine (SVM) and neural network (NN) to predict the energy consumption in the industries. They concluded that for the prediction of daily consumption, the neural network is more accurate than SVM (Grolinger et al., 2016). In the drilling process, feed rate, spindle speed, drill diameter of thrust force and torque are the critical parameters for drill wear. Backpropagation neural network is an efficient tool for predicting the effect of feed rate, spindle speed and drill diameter on drill wear (Singh et al., 2006).

7.4.4.4 Perspective Analytics

Perspective analytics is the highest level of data analytics. This analytics gives us the course of action to achieve the set goal. Perspective analytics requires more computational effort. As compared to the other analytics perspectives, perspective analytic is not foolproof. Before applying the perspective analytics, the target must decide. The hybridization of radial basis function network and non-dominated sorting genetic algorithm is an efficient tool for determining the critical parameter for efficient utilization of wire-cut electrical discharge machining (Saha et al., 2013).

7.4.5 Decision-Making

It is not possible to include all the parameters in the model; models are always built on sets of assumption, so it is very important to do qualitative analysis of models. All the suggestion of model cannot be implemented due to different physical constraints. So, manual decisions must be included.

7.4.6 Implementation

Implementing the new strategy is no easy task. A lot of brainstorming is required to implement new strategies. Whatever methods are decided, they must be implemented through proper processes.

7.4.7 Monitoring and Feedback

After successful impanation of strategies, it is very vital to monitor the status of the systems; also, there should be a feedback system for improvements.

7.5 CASE STUDY—LATHE MACHINE

To demonstrate the feasibility of implementing IoT in metal cutting, a lathe machine is used. Figure 7.4 shows that the system framework consists of a machine tool, industrial devices and computing units. The machine tool consists of various instruments, sensors and actuators connected by wires or wirelessly.

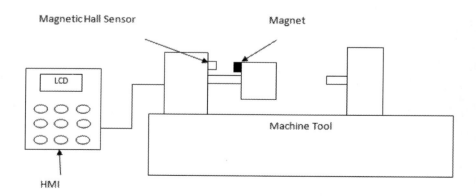

FIGURE 7.4 Machine tool HMI.

Road Map for IoT in Metal Cutting 127

The case study of the project-based industry is taken. The company carries out machining and fabrication of heavy large-size components. Due to the product's dynamic nature in terms of dimension and design, it is not easy to control the production activities. The industry has ten turning centers, four boring machines and three milling centers. The components need to be processed on more than one machine tool. In this industry, the project was delayed due to a bottleneck in a machine shop. Due to the uncertainty of demand and the high cost of machine tools, it is impossible to increase the number of machine tools. It is very important to improve productivity to complete the project on time. To find out the root cause of delay in machine shop, an IoT-based system is implemented to monitor value-added and non–value-added time.

7.6 MEASURING THE VALUE-ADDED TIME

In the machining process, when the actual manual cutting takes place is considered as value-added time. All other activities are considered as non–value-added time. It is challenging to measure real metal cutting time manually in machining processes, so an automated measuring system is developed to measure the actual metal cutting time using the magnetic hall sensor. On the back side of the chuck of the machine tool, a magnet is attached, and on the headstock cover, a magnetic sensor is attached. Whenever the magnet passes near to the magnetic hall sensor, a voltage change occurs in the sensor. If the chuck is rotating, there will continue to be changes in the voltage of the sensor. Thus, using the magnetic hall sensor data, we can find out if the chuck is rotating or steady. We can also find out the RPMs of the chuck using the frequency of change in voltage.

7.7 MEASURING THE NON–VALUE-ADDED TIME

Stoppages are classified into the following categories.

1. Setup time
2. Inevitable stoppages
3. Maintenance
4. Obligatory stoppages

7.7.1 SETUP TIME

Setup time is a period to prepare the machine tool for the machining. In project-based industries where frequent product changes occur, the monitoring of the setup time is essential. To calculate total setup time, we need to consider all the nonproductive stoppages like waiting for cranes, waiting for setup accessories and waiting for decisions, among others.

There is always the scope of reduction in the setup time in a project-based company. By controlling the setup time, we can increase the machine tool capacity.

To collect all these non–value-added times, the HMI (human–machine interface) is installed near the machine. On the HMI keypad is mounted an LCD screen.

Whenever the new components are started, the operator has to enter the specific code of that component through the HMI so that the total time taken to complete the component can be measured. The operator can input any number using the keypad. Each non–value-added activity has assigned a number that the operator has to press whenever he stops the machine. For example, if the machine has stopped for clamping the job during setup, the code is 01, where the first digit indicates setup and the latter digit represents sub activities.

7.7.2 INEVITABLE STOPPAGES

Stoppages that occur due to poor management of resources are considered nonproductive stoppages. The stoppages in Table 7.1 are considered inevitable stoppages.

7.7.3 MAINTENANCE

Machine tools are frequently stopped for various maintenance activities. There are five different types of maintenance performed on a machine tool, as follow.

1. Preventive maintenance
2. Planned maintenance
3. Predictive maintenance
4. Autonomous maintenance
5. Breakdown maintenance

7.7.3.1 Preventive Maintenance

Preventive maintenance is periodic maintenance to avoid sudden breakdown of equipment. The changing of the engine oil periodically in an automobile is a real-life example of preventive maintenance. A machine learning algorithm can predict the maintenance period (Li et al., 2021).

7.7.3.2 Planned Maintenance

Sometimes anomaly is detected in the equipment, and it is not required to stop the operation, but the operating cost is increased due to lower performance. Instead of doing immediate maintenance, maintenance is planned; such maintenance is called planned maintenance. For example, sometimes in an automobile, an anomaly in the engine fuel consumption leads to higher operating costs, but it does not stop the

---

TABLE 7.1
Inevitable Stoppages

1. Absenteeism	2. Waiting decisions	3. Waiting for a crane	4. Lack of a device
5. Lack of tool	6. Marking	7. Repair	8. Waiting for quality

Road Map for IoT in Metal Cutting

vehicle. The risk assessment in the planned maintenance is very important for the equipment's service life (Crognier et al., 2021).

7.7.3.3 Predictive Maintenance

In predictive maintenance, the prediction of the equipment's failure is carried out by examining the current condition of the equipment. For example, in the automobile, the period of changing the engine oil is predicted from the engine oil's appearance. Due to the development of sensors and artificial intelligence, the reliability of predictive maintenance is increased. The prediction of bearing failure can be made by sensor data like vibration, sound, current speed and temperature (Cakir et al., 2021)

7.7.3.4 Autonomous Maintenance

The equipment operator carries out autonomous maintenance daily before starting the equipment. This consists of checking lubricant level, noise level, pressure, etc. The maintenance departments provide a set of instructions to carry out the autonomous maintenance. Augmented reality can be a very efficient tool for training the operator for autonomous maintenance (Jia et al., 2021).

7.7.3.5 Breakdown Maintenance

The previously mentioned types of maintenance do not stop the machine from working, but in case of breakdown, maintenance equipment cannot perform the task, so immediate maintenance is required. Even if a machine is in operating condition despite the breakdown, it is not recommended to operate due to safety issues. For example, in the case of an automobile, the vehicle cannot be operated if there is a puncture in the wheel. If the vehicle brake fails, the vehicle can be operated but this is not recommended due to safety issues. Breakdown maintenance causes delays in manufacturing, so it is essential to avoid equipment breakdown. By using the previously mentioned maintenance techniques, reduction in breakdown is possible. The digital twin can be used to reduce the repair time during the equipment's breakdown (Aivaliotis et al., 2019).

7.7.4 OBLIGATORY STOPPAGES

Obligatory stoppages are those stoppages essential for metal cutting operation; without these stoppages, it is impossible to carry out the machining operation. The obligatory stoppages time can be minimized using sophisticated devices, but economic justification is required to use sophisticated devices. Table 7.2 lists the major obligatory stoppages in the metal cutting operation.

TABLE 7.2
Obligatory Stoppages

1. Change insert	2. Change tool holder/Support	3. Self-inspection	4. Measurement records
5. Self-inspection	6. Quality Inspection	7. Cleaning	8. Change turning to milling

TABLE 7.3
Stoppages Designation System

Activity	Code	Subactivity	Code	Final Code
Setup	0	Positioning	1	01
		Marking	2	02
		Clamping	3	03
		Unclamping	4	04
		Reference	5	05
		Drg/Doc reading	6	06
Inevitable	1	Absenteeism	1	11
stoppages		Waiting for a decision	2	12
		Waiting for a crane	3	13
		Lack of device	4	14
		Lack of tool	5	15
		Marking	6	16
		Repair	7	17
		Waiting for quality	8	18

7.8 SENSORS FOR QUALITY IMPROVEMENT AND CONDITIONING MONITORING

For quality improvement and conditioning monitoring, the following sensors are used.

1. CCD camera
2. Accelerometer
3. Strain gauge
4. Coolant flow meter

7.9 DATA APPLICATION

The collected data can be used by the following various departments.

7.9.1 PRODUCTION DEPARTMENT

The production department can use the collected data to analyze the various stoppages. It is found that the major stoppage of the machine tool is setup activates. The setup activity data of one of the project industries indicates that 25% of the time in setup activities is wasted in getting the device and tools.

Besides the main stops like setup, another vital source of waste is unavoidable stoppages like changing the cutting tool, self-inspection, cleaning, etc. The machine monitoring system can be used to identify the major obligatory stoppages, and corrective action can be taken.

Road Map for IoT in Metal Cutting

7.9.2 Planning Department

The current manufacturing environment has challenges like high complexity, dynamic production conditions and unstable market conditions. Also, there is a need for product customization with low cost and reducing the lead time (Schuh et al., 2017). To overcome these types of challenges, industries are looking for new technologies. IoT can be a superior technology to overcome the challenges in the current situation of the market. In the context of Industry 4.0, the function of production planning and control (PPC) can be forecasting the global quantities to be produced to increase the profitability, productivity and delivery time (Arnold and Chapman, 2004). It also involves the control of the manufacturing process, allowing real-time integration of resources (Moeuf et al., 2018). The integration of machine learning algorithms, real-time data collected with IoT, and production planning can help in the decision-making during the planning phase.

1. Prepare the capacity analysis for all process steps based on the cycle times and availability.
2. Study the process flow to identify the bottleneck machines, and classify the remaining machines as feeding machines and noncritical machines.
    1. Bottleneck machines are those that define the maximum capacity for the line or machining area. For these machines, the objective is to achieve the highest possible run time to maximize production capacity.
    2. Feeding machines are those that feed the bottleneck machines. For these machines, the objective is to produce parts following the planned sequence and timing in order to be able to feed the bottleneck machines in the right time. The timing shall be given by visual pull signals, based on the standard work in process in the lines.
    3. Noncritical machines are those located after the bottleneck machines in the process flow. For these machines, the objective is to keep the production flow to achieve an overall reduced lead time.
3. Planned hours are hours which are decided during the creation of a process for a component; actual hours are the ones which are taken from a runtime data analysis report.

In this context, prepare the comparison report between the planned hours to the actual hours taken in the shop for each component. As based on this comparison, the gap analysis between planned hours and actual hours must be done. During analysis, operator skills, machining tools, machining parameters and welding parameters should be considered. The objective is to bridge the gap and to reduce cycle time with respect to previous components through continuous improvement. Analysis has to be done to identify possibilities to reduce planned hours for critical components and actions taken for the same.

7.9.3 Quality Management

Quality management mainly consists of quality planning, quality control, quality assurance and quality improvements. Quality management consists of deciding quality objectives, operational processes and resources to fulfill the quality objective. Quality control focuses on fulfilling quality requirements. Quality assurance focuses on providing confidence that quality requirements will be fulfilled. Quality improvement seeks continuous improvement in product quality. Real-time data integration in quality management can improve the quality. The algorithm of perspective analytics is used for various operational processes to achieve the quality objective. The critical parameters for controlling the quality of the product can be decided using fuzzy logic or artificial neural networks, to name a two. Quality assurance is done by automatically stopping the machine tool if the operator is operating the machine tool on improper parameters. An optimization algorithm can be used to determine the optimum parameter for continuous improvement in product quality.

7.9.4 MAINTENANCE MANAGEMENT

The integration of IoT in maintenance systems can improve overall equipment efficiency. The diagnostic analytics algorithm helps determine the root cause of failure, so repair time came to be reduced. Real-time data like temperature and vibration, among others, can help in predictive maintenance.

7.10 CONCLUSION

This study presents an in-depth discussion on the implementation of IoT in metal cutting; it first presents the stages of IoT in metal cutting and discusses IoT's necessities and challenges in metal cutting. The enabling technologies of IoT in metal cutting are summarized according to critical activities of manufacturing: production, planning, quality management and maintenance. The study shows that the production department can identify the problem on the shop floor using real-time information. We can effectively measure the non–value-added and value-added activity in metal cutting processes discussed in detail. The planning department can compare the planned hours and actual hours, doing accurate planning for the next cycle. Also, identifying bottlenecks and feeding of machines helps the production department aware of on which machine more focus is required so that there will be on-time completion of components. The quality department can use IoT to decide the optimum parameters to improve the quality of the products. Maintenance departments can improve the OEE by real-time conditioning monitoring. Also, the collected data can be used for predictive maintenance using the various artificial intelligence algorithms.

7.11 LIMITATIONS OF THE STUDY

Cyber-security is the biggest challenge for Industry 4.0, which is not covered in this study. The industry's data management can crash due to unauthorized data changes, unauthorized deletion of data, and unauthorized access. Integration of design, marketing and supply chain is also not included in this study.

7.12 FUTURE SCOPE

In this study, the HMI is used to store the various activities of machining. However, due to human intervention, there is always the chance of mistakes. The research work is required to automatically record these activities to gather 100% accurate data. One of the possible solutions is to use image processing. Using the convolutional neural network, we can train the algorithm, and then it can be used for automatic recording of time.

REFERENCES

Aivaliotis, P., Georgoulias, K., & Alexopoulos, K. (2019). Using digital twin for maintenance applications in manufacturing: state of the Art and Gap analysis. In *2019 IEEE International Conference on Engineering, Technology and Innovation (ICE/ITMC)* (pp. 1–5). IEEE, France.

Alptekinoğlu, A., & Corbett, C. J. (2008). Mass customization vs. mass production: variety and price competition. *Manufacturing & Service Operations Management*, 10(2), 204–217.

Alvian, P., Eun, Y., Meerkov, S. M., & Zhang, L. (2020). Smart production systems: automating decision-making in manufacturing environment. *International Journal of Production Research*, 58(3), 828–845.

Arnold, J. T., & Chapman, S. N. (2004). *Introduction to materials management*. Pearson Education India.

Athreya, A. P., DeBruhl, B., & Tague, P. (2013, October). Designing for self-configuration and self-adaptation in the Internet of things. In *9th IEEE International Conference on Collaborative Computing: Networking, Applications and Worksharing* (pp. 585–592). IEEE, USA.

Atzoria, L., Iera, A. V., & Morabito, G. (2010). The internet of things: a survey. *Computer Networks*, 54(15), 2787–2805.

Benardos, P. G., & Vosniakos, G. C. (2017). Internet of things and industrial applications for precision machining. In *Solid state phenomena* (Vol. 261, pp. 440–447). Switzerland: Trans Tech.

Cakir, M., Guvenc, M. A., & Mistikoglu, S. (2021). The experimental application of popular machine learning algorithms on predictive maintenance and the design of IIoT based condition monitoring system. *Computers & Industrial Engineering*, 151, 106948.

Chen, X., Li, C., Tang, Y., Li, L., & Xiao, Q. (2018). A framework for energy monitoring of machining workshops based on IoT. *Procedia CIRP*, 72, 1386–1391.

Cooper, C., & Kaplinsky, R. (Eds.). (2005). *Technology and development in the third industrial revolution*. London: Routledge.

Crognier, G., Tournebise, P., Ruiz, M., & Panciatici, P. (2021). Grid operation-based outage maintenance planning. *Electric Power Systems Research*, 190, 106682.

Dai, H. N., Wang, H., Xu, G., Wan, J., & Imran, M. (2020). Big data analytics for manufacturing internet of things: opportunities, challenges and enabling technologies. *Enterprise Information Systems*, 14(9–10), 1279–1303.

Deshayes, L., Welsch, L., Donmez, A., Ivester, R., Gilsinn, D., Rhorer, R., . . . & Potra, F. (2006). Smart machining systems: issues and research trends. In *Innovation in life cycle engineering and sustainable development* (pp. 363–380). Dordrecht: Springer.

Dutta, S., Pal, S. K., Mukhopadhyay, S., & Sen, R. (2013). Application of digital image processing in tool condition monitoring: a review. *CIRP Journal of Manufacturing Science and Technology*, 6(3), 212–232.

Frader, L. L. (2005). *The industrial revolution: a history in documents*. Oxford: Oxford University Press.

Grolinger, K., L'Heureux, A., Capretz, M. A., & Seewald, L. (2016). Energy forecasting for event venues: big data and prediction accuracy. *Energy and Buildings*, 112, 222–233.

Guoqiang, S., Yanming, C., Chao, Z., & Yanxu, Z. (2013). Design and implementation of a smart IoT gateway. In *2013 IEEE International Conference on Green Computing and Communications and IEEE Internet of Things and IEEE Cyber, Physical and Social Computing* (pp. 720–723). IEEE, Beijing, China.

Hentz, J. B., Nguyen, V. K., Maeder, W., Panarese, D., Gunnink, J. W., Gontarz, A., . . . & Hascoët, J. Y. (2013). An enabling digital foundation towards smart machining. *Procedia Cirp*, 12, 240–245.

Jerard, R. B., Fussell, B. K., Suprock, C. A., Cui, Y., Nichols, J., Hassan, R. Z., & Esterling, D. (2009, January). Integration of wireless sensors and models for a smart machining system. In *International Manufacturing Science and Engineering Conference, ASME* (Vol. 43628, pp. 119–128), West Lafayette, Indiana, USA.

Jia, W., Zhu, J., Xie, L., & Yu, C. (2021). Review of the Research on augmented reality maintenance assistant system of mechanical system. In *Journal of Physics: Conference Series* (Vol. 1748, No. 6, p. 062041). IOP Publishing, UK.

Kamble, S. S., Gunasekaran, A., Ghadge, A., & Raut, R. (2020). A performance measurement system for industry 4.0 enabled smart manufacturing system in SMMEs-A review and empirical investigation. *International Journal of Production Economics*, 229, 107853.

Khalifa, O. O., Densibali, A., & Faris, W. (2006). Image processing for chatter identification in machining processes. *The International Journal of Advanced Manufacturing Technology*, 31(5), 443–449.

Kim, D. H., Kim, T. J., Wang, X., Kim, M., Quan, Y. J., Oh, J. W., . . . & Ahn, S. H. (2018). Smart machining process using machine learning: a review and perspective on machining industry. *International Journal of Precision Engineering and Manufacturing-Green Technology*, 5(4), 555–568.

Kozjek, D., Kralj, D., & Butala, P. (2017). Interpretative identification of the faulty conditions in a cyclic manufacturing process. *Journal of Manufacturing Systems*, 43, 214–224.

Lee, Y., & Dornfeld, D. A. (1998). Application of open architecture control system in precision machining. In *31st CIRP International Seminar on Manufacturing Systems* (pp. 436–441), University of California, Berkeley.

Li, L., Wang, Y., & Lin, K. Y. (2021). Preventive maintenance scheduling optimization based on opportunistic production-maintenance synchronization. *Journal of Intelligent Manufacturing*, 32(2), 545–558.

Liang, S. Y., Hecker, R. L., & Landers, R. G. (2004). Machining process monitoring and control: the state-of-the-art. *The Journal of Manufacturing Science and Engineering*, 126(2), 297–310.

Moeuf, A., Pellerin, R., Lamouri, S., Tamayo-Giraldo, S., & Barbaray, R. (2018). The industrial management of SMEs in the era of Industry 4.0. *International Journal of Production Research*, 56(3), 1118–1136.

Park, H. S., Qi, B., Dang, D. V., & Park, D. Y. (2018). Development of smart machining system for optimizing feedrates to minimize machining time. *Journal of Computational Design and Engineering*, 5(3), 299–304.

Park, H. S., & Tran, N. H. (2014). Development of a smart machining system using self-optimizing control. *The International Journal of Advanced Manufacturing Technology*, 74(9–12), 1365–1380.

Patra, K., Pal, S. K., & Bhattacharyya, K. (2013). Strategies for intelligent drill wear prediction using multiple sensor signals. *International Journal of Mechatronics and Manufacturing Systems 23*, 6(5–6), 493–512.

Perzylo, A., Kessler, I., Profanter, S., & Rickert, M. (2020). Toward a knowledge-based data backbone for seamless digital engineering in smart factories. In *2020 25th IEEE International Conference on Emerging Technologies and Factory Automation (ETFA)* (Vol. 1, pp. 164–171). IEEE, Austria.

Saha, P., Tarafdar, D., Pal, S. K., Saha, P., Srivastava, A. K., & Das, K. (2013). Multi-objective optimization in wire-electro-discharge machining of TiC reinforced composite through Neuro-Genetic technique. *Applied Soft Computing*, 13(4), 2065–2074.

Saqlain, M., Piao, M., Shim, Y., & Lee, J. Y. (2019). Framework of an IoT-based industrial data management for smart manufacturing. *Journal of Sensor and Actuator Networks*, 8(2), 25.

Schuh, G., Reuter, C., Prote, J. P., Brambring, F., & Ays, J. (2017). Increasing data integrity for improving decision making in production planning and control. *CIRP Annals*, 66(1), 425–428.

Sharma, R., Saini, A., & Rana, G. (2021a). Big data analytics and businesses in Industry 4.0. *Design Engineering*, 2021(2), 238–252.

Sharma, R., Saini, R., Prakash, C., & Prasad, V. (2021b). *Internet of things and businesses in a disruptive economy*, 1st ed. New York: Nova Science Publishers.

Singh, A. K., Panda, S. S., Chakraborty, D., & Pal, S. K. (2006). Predicting drill wear using an artificial neural network. *The International Journal of Advanced Manufacturing Technology*, 28(5), 456–462.

Trivedi, J. P. (2015). Mobile advertising effectiveness on Gen Ys attitude and purchase intentions'. *International Journal of Marketing and Business Communication*, 4(2), 2.

Urbikain, G., & de Lacalle, L. L. (2020). MoniThor: a complete monitoring tool for machining data acquisition based on FPGA programming. *SoftwareX*, 11, 100387.

Venkataraman, K., Ramnath, B. V., Kumar, V. M., & Elanchezhian, C. (2014). Application of value stream mapping for reduction of cycle time in a machining process. *Procedia Materials Science*, 6, 1187–1196.

Vinod, V., & Sridharan, R. (2008). Scheduling a dynamic job shop production system with sequence-dependent setups: an experimental study. *Robotics and Computer-Integrated Manufacturing*, 24(3), 435–449.

Wang, C., Ghani, S. B., Cheng, K., & Rakowski, R. (2013). Adaptive smart machining based on using constant cutting force and a smart cutting tool. *Proceedings of the Institution of Mechanical Engineers, Part B: Journal of Engineering Manufacture*, 227(2), 249–253.

Wang, Z., Ritou, M., Da Cunha, C., & Furet, B. (2020). Contextual classification for smart machining based on unsupervised machine learning by Gaussian mixture model. *International Journal of Computer Integrated Manufacturing*, 33(10–11), 1042–1054.

Yu, J., Zheng, X., & Wang, S. (2019). A deep autoencoder feature learning method for process pattern recognition. *Journal of Process Control*, 79, 1–15.

Zhang, F., Liu, M., Zhou, Z., & Shen, W. (2016). An IoT-based online monitoring system for continuous steel casting. *IEEE Internet of Things Journal*, 3(6), 1355–1363.

Zhang, Y., Qu, T., Ho, O. K., & Huang, G. Q. (2011). Agent-based smart gateway for RFID-enabled real-time wireless manufacturing. *International Journal of Production Research*, 49(5), 1337–1352.

8 Reinventing HR with Conversational Artificial Intelligence
A Conceptual Framework

Neetu Kumari and Geeta Rana

CONTENTS

8.1 Introduction 137
8.2 Literature Review 139
8.3 Objectives 140
8.4 Research Methodology 140
8.5 Conventional Human Resource Management 140
8.6 Conceptual Model of Artificial Intelligence and Human
     Resource Management 142
     8.6.1 Managerial Function—Artificial Intelligence HR
          Decision-Making System 143
     8.6.2 Procurement—Artificial Intelligence Procurement 143
     8.6.3 Training and Development—Artificial Intelligence
          HR Training and Development 145
     8.6.4 Compensation—Artificial Intelligence Incentive System 145
     8.6.5 Maintenance—Artificial Intelligence HR Maintenance 146
     8.6.6 Integration—Artificial Intelligence HR Integration 146
     8.6.7 Emerging Issue—Artificial Intelligence HR Systems 146
8.7 Discussion and Managerial Implications 146
8.8 Conclusion 148
References 149

8.1 INTRODUCTION

Over the past few years, artificial intelligence (AI) has become an integral part of our daily lives and management. It was such an extraordinary move that gathered the attention of the people from all the corners of the world. The companies are moving from gigantic data to machine learning and further to artificial intelligence (AI) (Tambe et al., 2018; Rana & Sharma, 2019). Artificial intelligence can be defined as a programmed algorithm that has a learning capacity and aims to be like human but surpasses humans in its ability. In other words, it is an advanced technology that allows a computer to perform all those errands that generally require human

DOI: 10.1201/9781003145011-8

137

cognition. It is a preliminary definition that captures some of the important characteristics of AI, but over a period of time, its application area has extended and it has evolved to a much more diversified area. According to Kaplan and Haenlein (2019), AI is a system's knack to appropriately understand huge data use its learnings to achieve explicit goals and tasks.

Gene engineering, cloud computing, IoT, biotechnology and 3D printers are the products of new technological revolution, and AI plays a vital role in integrating all these technologies. Hyper connectivity has distinguished new digital technology from preceding technologies, and AI is a connecting link between the two. For example: Big Data needs to be analyzed by AI and robots, and machines require AI to function efficiently and thus represent digital transformation.

AI has infiltrate in management and have impact on the employees' work in human resource management. AI is the capability of machines to emulate human intelligence, and also allows the machines to learn automatically and adapt it in order to provide refined responses to the situation. But the question is: How can AI can be used in human resources? In an HR department that works with the human element, it is important to see how companies use AI to provide employees with enhanced experiences (Durrani, 2020).

In today's competitive world, AI has influenced almost everything, and HR departments are among them. Human resource management has understood the necessity of integrating human knowledge with machine learning for an effortless work and predictive work environment. AI is such an innovation that has shaped real-time decision-making that follows a definite set of innovative computing techniques and algorithms. By amalgamating human resource departments with AI, the experience of workforces would be enhanced. AI plays an important role in understanding the mindset of its target audience and strategizes a plan accordingly to achieve the set goals. Today, with the help of AI, machines have become capable of imitating human intelligence. AI utilizes the collected data in order to provide data-driven decisions. Thus, in this chapter, we will study the impact of artificial intelligence in HR decision-making strategies.

HR professionals have realized the importance of integrating the human mind with machine learning for an effortless workflow which has been proved by the mentioned reviewed literature. A large number of companies are using AI for their daily work schedules, and some of them have even developed their own AI solutions. Artificial intelligence has been around for decades, but recently its application has become more prominent due to advance technological development. Still, the use of AI technology is relatively less because of cost–benefit analysis. AI is used in almost every sector these days like health, education, entertainment, manufacturing, etc., but this chapter highlights the applicability of AI in human resources.

This chapter encompasses a review of literature on conventional HRM and their functions. It also highlights AI and its use in HRM. The chapter concludes with discussion and managerial implications. The study suggests various recommendations for companies and individuals that help them to involve AI in their daily routine.

8.2 LITERATURE REVIEW

The reviewed literature states that intelligent automation is a new methodology to manage an organization's employees and it enhances a firm's output and offers new way of development of HRM, but also leads to challenges at ethical and technological levels. The study also found positive impact of technology on HRM functioning like human–robot collaboration, job replacement, learning and decision-making, etc. (Vrontis et al., 2021).

Prentice, Dominique Lopes and Wang (2020) found that service experience with employees and AI significantly related to customer loyalty and engagement. Other researchers (Johnson et al., 2020) found that e-HRM and AI have the impact to change how recruiting and selecting candidates in procurement of employees takes place in the hospitality and tourism industry. But it must be confirmed that the decision must be cleared to employees, which leads to better organizational outcomes. Pillai and Sivathanu (2020) worked on the application of AI technology for talent acquisition. The results reveal that top authority support, HR alertness, competition, cost management and support from AI provides significantly and positively influenced application of AI technology. The study also highlighted that traditional methods of talent acquisition negatively moderates the relationship between adoption of AI technology actual wastage.

Yawalkar (2019) examined the role and challenges of artificial intelligence in human resource practices. The various functions performed by companies like recruitment and hiring of employees, data collection and analyzing, workforce and work management, reducing workload and increasing efficiency of employees in the workplace can be easily done with the help of robotics intelligence Rana and Sharma (2019).

Jain (2018) identified the role of AI in HRM and examined the adoption of modern techniques in various HR processes like recruitment, performance appraisal, cloud-based HR systems, etc. AI in recruitment process helped in screening the candidates, automatically generating messages to them, employee dealings, arranging interviews, etc. (Ahmed, 2018). AI is a system's ability to learn from data and to use data to achieve organizational goals. Merlin and Jayam (2018) highlighted the use of AI in workplace and how it can be a great help for HR professionals in understanding their work, identifying the problems attached to it and trends in advance. The AI for human resources is useful in decision-making and dealing with uncertainty (Jarrahi, 2018). According to Geetha and Reddy (2018), the role of human resource in organization is quite important and the working of all other technologies depends on human efforts. Jain (2018) identified the role of AI in HRM and examined the adoption of modern techniques in various HR process like recruitment, performance appraisal, cloud-based HR systems, etc. AI's role is quite important in the whole employment process (Ahmed, 2018).

AI technologies in HR development was discussed by Buzko et al. (2016), who contemplated the various hurdles in path of application of AI technology in HR.

According to Wang and Kosinski (2018), AI is making machines intelligent and intelligence enables an entity to perform properly. Dirican (2015) wrote that the use of AI and robotics in organizations can have a negative impact on their overall functioning.

Budhwar and Debrah (2001) highlights rapid development of human resource management practices in a fast-growing world and also focused on the need for having cross-national HRM practices to perform HRM functions effectively globally. The role of artificial intelligence and its use for human resource management has also been examined by Kapoor (2010).

Tzafrir (2007) analyzed the trust mechanism and organizational performance, and how HR practices impact it. The study found that HR managers work on internal promotion systems and provide conducive training programs if the trust lies in organization. The study further reveals that trust also positively impacts organizational performance.

8.3 OBJECTIVES

The integral part of any organization is its human resources. One can come across effective HR practices that can improve performance. Among the numerous HR innovative practices, one is artificial intelligence. Nowadays, it is necessary skillset for employees to involve AI in human resources departments. The first thing that HR managers should do is involve AI in HR functions, because it will influence employee performance. The reviewed literature reveals that there is a still lack of overall AIHRM framework which includes all the specific HRM dimensions. The objective of the study is to study the conventional HRM model and propose the conceptual AI human resource management model which provides theoretical and conceptual guidelines for future researchers.

8.4 RESEARCH METHODOLOGY

To achieve the purpose of the study, we have used descriptive research design and reviewed diverse types of literature reviews. As the study is about HRM and its functions to propose the conceptual model, secondary data is used. We utilized electronic databases for literature searches. The secondary data has been collected from HR journals, blogs, textbooks, research papers, survey reports, online human resource websites and other related publications, both online and offline.

8.5 CONVENTIONAL HUMAN RESOURCE MANAGEMENT

In this section, we will discuss conventional human resource management and its functioning discussed by Rao (2013). All kind of organization whether small and big are focusing on HRM. The HRM process begins with two dimensions: managerial functions and operative functions.

1. **Managerial functions**: Management is basically personnel administration. It is the development of people working in an organization. The basic managerial functions comprise PODC (i.e. planning, organizing, directing and controlling).

Reinventing HR with Conversational AI

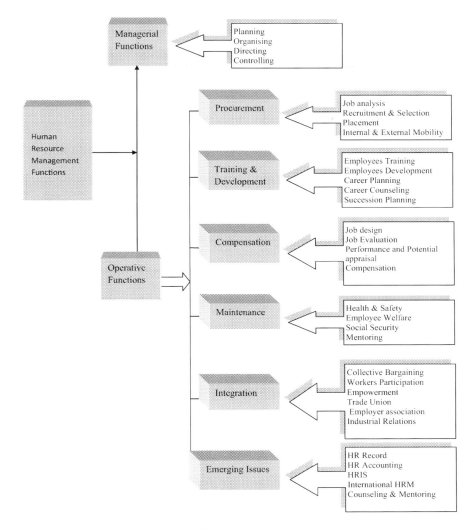

FIGURE 8.1 Conventional model of functions of human resource management.

Planning: Planning is the first important step of any successful enterprise. This function deals with the plans, programs, strategies, procedures and policies to achieve the desired results. It includes upcoming course of action to reach adequate results.

Organizing: After planning for the future, the next step is to allocate tasks and responsibilities among employees. This includes identifying employee activities, assigning duties, delegating authority and coordinating activities of individuals and groups.

Directing: Directing involves getting employees together, commanding or motivating them to do work effectively to achieve organizational goals.

It includes tasks of executing plans, reviewing employees safety measures, compensation, discipline, etc. It is the essential for any organization; without direction, it cannot reach its destination.

Controlling: Controlling includes checking, verifying and regulating whether everything done according to plan. Controlling function of management comprises reviewing personnel policies, programs, job evaluation, measuring the employee's performance, training and employee development.

2. **Operative functions:** The pervasive functions include various key performance areas like procurement, development, motivation, compensation, maintenance, integration and emerging issues.

Procurement function: It is the first important function of human resource management. It includes procuring people who possess required skills, knowledge and aptitude. It includes manpower required, recruiting and selecting candidates, job analysis (job description and specification), placement and internal mobility (transfers, layoffs, discharge, separation).

Development: This function considers developing skills, knowledge, attitudes and creative ability, and improving employees' performance with training. It is the process of drafting and directing training for all employees based on employee potential and performance.

Motivation and compensation : This includes provides adequate and equitable remuneration to all working in an organization. The fair remuneration motivates employees to give their best to the organization. They are facilitated with both intrinsic and extrinsic rewards.

Maintenance: This deals with sustaining and improving the psychological and physical health of employees. It aims at caring and preserving the people working in organization through a variety of welfare programs.

Integration function: Managerial activities that integrate human resource and organization through various employee-oriented programs are called integration. It includes activities like grievance redressal, discipline, teamwork, collective bargaining, employee participation and industrial relations.

Emerging Issues: Effective human resource management practice depends on refining human resource practices to changing conditions. Some important issues affecting HR practice nowadays are personnel records, human resource audits, human resource information systems, human resource research, human resource accounting and international HRM. All these functions keep employee motivated to provide their best in a dynamic and ever-changing human resource practice.

8.6 CONCEPTUAL MODEL OF ARTIFICIAL INTELLIGENCE AND HUMAN RESOURCE MANAGEMENT

AI combines human resource functions with automatic technology which requires a deep indulgent by teams involved (Meister, 2019). AI depends on human data to work; it depends on data we feed. If the data input is wrong, the result will also include errors. For example, if the information of age of HR would not be included in the data, it would be not possible to detect this information (Zielinski, 2017). In this chapter, we consider modern HRM processes by including modern HRM that differs from conventional HRM. With the technological advancement, the HR working environment is also changing. The conventional HR approach may not be efficient today, including with AI technology improving it significantly. The proposed model discusses how artificial intelligence inclusion will change HR. All the statistical data used in human resources—for example, HR planning, procurement, training, development, compensation, performance appraisal, etc. are fed to the AI machine to make HR managerial functions effective (Yano, 2017). The AI intervention will reduce worker hours in tasks and the reduced time period can be used by manager to engage workers in useful tasks.

The proposed conceptual model helps human resource managers to take timely and effective decisions. The framework considers how AI can be joined with human resource management practices. The application of AI in HRM, forming an artificial intelligence human resource system (AIHRM), is described and explained in Figure 8.2.

8.6.1 Managerial Function—Artificial Intelligence HR Decision-Making System

Human resource managerial function decision-making–including planning, organizing, staffing and directing–is the starting point of HRM. AI technology helps managers as a supporting decision-making system. Technology such as knowledge discovery, data mining, etc., help in collecting global information and taking perfect planning, organizing, directing and controlling. AI will diminish idleness of workers at places of work because numerous tasks are carried out by robotic technology which increases the efficiency at workplaces. Robotic technology helps in identifying required data from available data, filing reports, copying data, processing, collecting data for payroll systems, etc. (Yawalkar, 2019).

8.6.2 Procurement—Artificial Intelligence Procurement

The prevalent area of HR where AI technology is at present used is procurement. The second important function of HR is procurement, including screening résumés, interviewing and selection. With the intervention of AI technology and Big Data methods in recruitment and selection procedures, it is quite easy to screen suitable candidates from among thousands of applicants without bias (Prakash et al., 2021). AI technology can be used in recruitment via chatbots. Many startup companies are using these bots in their recruitment processes. AI chatbots appear on company websites, greet the possible candidates and make them aware about the open positions in the company. The chatbots help in the pre selection process by determining whether an applicant could be suitable for the position. It analyzes candidates, ascertaining qualifications, skills and knowledge, suitability and responses in interviews. It helps screen candidates on the basis of scores and qualification, and rates them (Ideal, 2019). In addition, natural language processing helps in conversion from speech to text which will reduce recruiters' work. A selection process applicant tracking

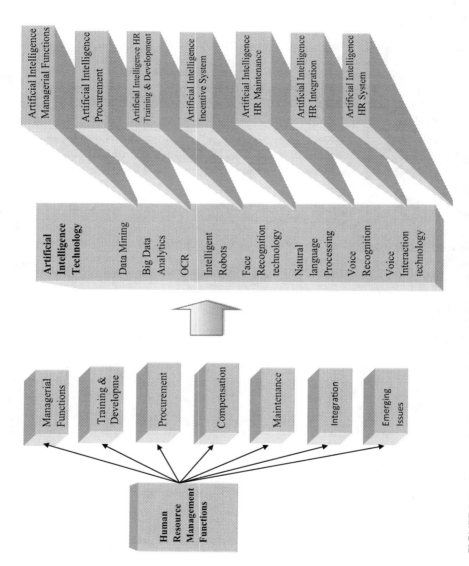

FIGURE 8.2 A conceptual framework of artificial intelligence human resource management.

Reinventing HR with Conversational AI

system (ATS) can be used to help scan candidates' CVs and searching keywords (Lee, 2019). The AI intervention reduces recruitment and selection costs and time. One more important technology that can be used to analyze candidates and can suggest suitable positions to candidates to match their profiles is optical character recognition (OCR) technology (Hutson, 2017). Facial recognition techniques are extremely useful in determine that a candidate is consistent with the documentation provided (Jain & Li, 2011). The AI technology voice test method can easily detect facial expression, choice of words, confidence and knowledge of interviewer, and help the recruiter select the most suitable candidate. It also helps in internal mobility of employees by providing learning opportunities.

8.6.3 TRAINING AND DEVELOPMENT—ARTIFICIAL INTELLIGENCE HR TRAINING AND DEVELOPMENT

In an organization, incessant training and development of employees are important steps. Both new employees and underperforming employees need training throughout their careers to effectively perform their responsibilities (Rana & Goel, 2015). In the training process, every employee is different from others, and might have different learning tactics (Sharma et al., 2019); here, AI helps to craft the effective decisions. The machine can identify the areas the staff need help by analyzing input data. According to Palmu (2020), AI creates customized learning programs for different stakeholders having different cultures, generations, education, personalities and working backgrounds. AI also help trainers to create good teaching materials and provide on-the-job training and offline training aids. AI can also be used in training program by chatbots. These chatbots are taught to answer trainee queries and provide answers instantly. The frequently asked questions during the training can be included in training teaching material, which reduces the amount of additional queries from trainees. The AI intervention helps trainer in attending each trainee personally, which helps the trainer to focus on important areas of their work.

8.6.4 COMPENSATION—ARTIFICIAL INTELLIGENCE INCENTIVE SYSTEM

AI applications help to facilitate fairness of incentive payments. The AI technology known as back propagation (BP) neural network can be used in designing financial and non-financial incentives to provide fair wage and salary administration. The process of job analysis like collecting data about former workers becomes quite easy using intervention of AI. According to Peter Hogg (2019), recruiters should consider the potential appraisal of the candidates over experience. AI systems help HR managers to analyze employees based on their performance reviews. The employees get more opportunity for learning, good remuneration, promotion and training. With AI work scheduling key metrics like experience, performances are analyzed automatically to generate effective labor schedules for maximum productivity. Analyzing real-time and historical operational data reduces over scheduling and under scheduling of work.

8.6.5 Maintenance—Artificial Intelligence HR Maintenance

The AI provides data for evaluating employees' performance, so employee maintenance can be done more frequently and more easily. It is easy to interact with the workers during training. The common bias in employee performance evaluation is reduced because it is evaluated by machine. AI is now being used to identify behaviors like stress and poor performance, and alert HR managers about employee performance variables that cause poor work performance. AI can also identify employees' attitude and behavior at the workplace that leads to accidents. The chatbots can make interactions intelligent and lead to employee self-service and candidate management. AI points out the "red flags" to HR managers by accessing organizational network data like e-mail traffic, comments, ethical lapses and acquiescence risk, so they can intervene before creating big problems in department.

8.6.6 Integration—Artificial Intelligence HR Integration

Digital technology is also playing important role in HR integration. The transparent and customized operation in the workplace through AI interventions improves employees' loyalty. The performance of employees can be easily evaluated without bias; they are given chances for career advancement inside the company. It also helps in potential appraisal and detects candidates for promotions and other rewards. Employee morale and performance is improved. The employees' external mobility will be reduced, and employee advocacy will increase. The company needs not to waste spending on hiring and training new employees, because current ones will not leave the company. AI technology acts as assistant, performer and consultant, and can be used to solve many HR problems like stereotypes, employee management, etc. (Jia et al., 2018). AI gives companies both tangible intangible benefits.

8.6.7 Emerging Issue—Artificial Intelligence HR Systems

AI makes the HR manager's task easier and less time consuming, reducing their workload. AI-enabled systems automatically respond to employees' queries and manage the work. Automating low-cost, less repeatable tasks gives HR managers time to help with the most important administrative work. The use of a "virtual assistant system" helps in composing emails, organizing meetings, coordinating participants and managing calendars; through this, it will gradually collect particulars and expand service areas by gripping the experience of employee and their contemporaries (Jia et al., 2018). AI-based software plays an important role in HR strategy; it helps in employee management, human resource accounting, HRIS, managing payrolls, revising company policies and practices, investigating company compliance, creating strategies, etc. (Jacobs, 2019).

8.7 DISCUSSION AND MANAGERIAL IMPLICATIONS

The intervention of AI has brought positive aspects to business and our daily lives, though a lot of discussions have been undertaken related to it. Initially, skepticism

Reinventing HR with Conversational AI

among the people is always caused whenever a new technology is introduced anywhere, and the same occurred with AI. Use of AI requires managing a lot of data to create information that can be risky. If the input data is not correct, the result will be also be inaccurate, leading to problems in the organization and correcting those leads to waste of time and money. Since for many organizations, this is comparatively a newer technology—and not every employee knows how to use it properly, and thus how to acquire it—there is a need to equip the staff with adequate training, which is expensive. These factors hinder startups and small businesses from using such kinds of technology properly, but the time is not far away when AI will be part of our daily routine; it will become less expensive—and much easier, as well. Meanwhile, large organizations are able to inculcate this new technology very quickly. According to Pribanic (2018), those organizations who employ AI perform better than their competitors having no AI.

AI has improved recruitment in many ways, but there are certain points to be considered here. Candidate need to enter their online applications in certain prescribed formats and only then will the machine be able to analyze it; therefore, those applications which are done face to face or in a paper form fall out of AI's radar. AI in recruitment can be negatively perceived by job seekers, as well, as sometimes resume might not be looked over by a person because they do not include specific keywords–but if everyone uses the same phrase or words in their resume, it will lose their credibility and will waste the time of both recruiter and candidates. In this regard, Lee (2019) said that the job seekers do not realize that they are being evaluated by a machine. According to them, not much effort is wasted on seeing the resume, while the decision has to be taken based on certain keywords.

Marr (2017) discussed managers ignoring the results of performance reviews if they are unable to go with belief of the employees. The machines are not able to make a fair judgment, as it lacks human emotions.

Though AI is used to assess the employee's performance, it will not take place of the employer–employee professional relationship. The machines only give the reviews, but the final decision is made by the managers who determine if the employee genuinely deserves the job or the promotion. The machine does not consider the employee's keenness to work, qualities, personality traits, etc.; it is only the manager who can evaluate such attributes. Privacy is another main area of concern as far as AI is concerned; it is questionable whether it is out of harm's way to share such huge data and information about particular employees' personal information via machine. AI engineers are working on making machine data secure. The European Union's General Data Protection Regulation (GDPR) was crafted in order to safeguard the private information of the populace. It is quite possible that any outsider may hack the system and can use the employees' personal information for their own benefit, and that can cause a huge problem for its employees. One more area of concern is that AI systems can collect information from anywhere and any voice tone, and it is also possible that it can collect information by spying on employees' private emails. This shows that AI technology can be misused.

When it comes to using technology, differences among generations is another key concern in AI. Those who resist it will find it difficult to work with AI-powered machines. Younger employees are usually more comfortable with modern technology that old ones, and this is more problematic for employees who are close to retirement. Young people would prefer to work with AI, whereas senior people feel happy to work face to face with their customers and colleagues (Bean, 2018).

AI has eliminated many jobs, and it will continue to do so in the future. In the coming years, many people will lose their jobs if they will be not retrained. In upcoming years, there are chances of joblessness, and people need to re-educate themselves to find new jobs.

One more issue in recruiting has come up; i.e. bias against some demographic groups by machines by not giving equal consideration to all applicants. It is found that AI recruitment algorithms find bias against one gender (female). Meyer (2018) stated that inspite of all the efforts in fixing it, the results always came back rejecting talented female candidates. The HR and IT companies must work together to secure and effective HR systems which will keeps employees' data secure.

8.8 CONCLUSION

Managing incessant advance is a big challenge in front of all organizations. To augment continuous growth, most organizations are focusing on adapting contemporary technologies. In this process, adoption of artificial intelligence is taking lead in all types of organizations. Many organization are using AI, which is playing an integral role in managerial functions. The study proposed a seven-part AI framework for HRM.

The human resource functions—i.e. managerial functions, training and development, procurement, compensation, maintenance, integration and emerging issues—when combined with AI technology—i.e. data mining, Big Data analytics, optical character recognition, intelligent robots, facial recognition technology, natural language processing, voice recognition, voice interaction technology, and scanning systems—will lead to AIHRM that includes AI HR decision-making systems, AI HR training and development, AI procurement, AI incentive systems, AI HR maintenance, AI HR integration and AI HR systems. This AI framework provides suggestions and managerial directions of AI implementation in HRM, and helps individuals and companies to transit from conventional HRM to a technical HRM whereby the use of AI is incorporated. The issues and concerns discussed in the study guide current issues related to the subject. More companies are adding AI into their operations, and it is becoming part of daily life. AI is successfully used in recruitment, selection, training and development processes. AI technology has made it possible to tailor customized training to employees according to extreme situations without harming long-term customer relationships. Employers are using AI successfully to evaluate employees' performance. The practice of AI technology definitely improves individual and organizational performance. We hope that the conceptual and practical findings of this chapter will move AI management of human resources frontward in appropriateness and efficiency. To prepare HR for future courses of events, HR managers should take the necessary actions to learn AI trends and lay a strong base for professional growth.

REFERENCES

Ahmed, O. (2018). Artificial Intelligence in HR. *International Journal of Research and Analytical Reviews*, 5 (4), 971–978.

Bean, S. (2018). *Generations Divide on the Role of Artificial Intelligence in the Workplace.* Retrieved from https://workplaceinsight.net/generations-divide-on-the-role-of-artificial-intelligence-in-the-workplace/ (Accessed 31st March 2021).

Budhwar, P. S. & Debrah, Y. (2001). Rethinking Comparative and Cross National Human Resource Management Research. *The International Journal of Human Resource Management*, 12 (3), 497–515.

Buzko, I., Dyachenko, Y., Petrova, M., Nenkov, N., Tuleninova, D. & Koeva, K. (2016). Artificial Intelligence Technology in Human Resource Development. *Computer Modelling & New Technologies*, 20 (2), 26–29.

Dirican, C. (2015). The Impacts of Robotics, Artificial Intelligence on Business and Economics. *Procedia—Social and Behavioural Sciences*, 195, 564–573.

Durrani, K. (2020). *The Impact of AI in Human Resource Decision Making Processes.* Retrieved from www.hrtechnologist.com/articles/ai-in-hr/the-impact-of-ai-in-human-resource-decisionmaking-processes/ (Accessed 20th March 2021).

Geetha, R. & Bhanu Sree Reddy, D. (2018). Recruitment Through Artificial Intelligence: A Conceptual Study. *International Journal of Mechanical Engineering and Technology*, 9 (7), 63–70.

Hogg, P. (2019). Artificial Intelligence: HR Friend or Foe? *Strategic HR Review*, 18 (2), 47–51.

Hutson, M. (2017). *Even Artificial Intelligence Can Acquire Biases against Race and Gender.* Retrieved from www.sciencemag.org/news/2017/04/even-artificial-intelligence-can-acquire-biases-against-race-and-gender (Accessed 22nd March 2021).

Ideal. (2019). *AI for Recruiting: A Definite Guide for HR Professionals.* Retrieved from https://ideal.com/ai-recruiting/ (Accessed 22nd March 2021).

Jacobs, E. (2019). How Artificial Intelligence Helps Companies Recruit Talented Staff. *Financial Times.* Retrieved from www.ft.com/content/2731709c-3043-11e9-8744-e7016697f225 (Accessed 22nd March 2021).

Jain, A. K. & Li, S. Z. (2011). *Handbook of Face Recognition.* New York: Springer.

Jain, D. S. (2018). Human Resource Management and Artificial Intelligence. *International Journal of Management and Social Sciences Research*, 7 (3), 56–59.

Jarrahi, M. H. (2018). Artificial Intelligence and the Future of Work: Human—AI Symbiosis in Organizational Decision Making. *Business Horizons*, 61 (4), 1–10.

Jia, Q., Guo, Y., Li, R., Li, Y. & Chen, Y. (2018). A Conceptual Artificial Intelligence Application Framework in Human Resource Management. In *Proceedings of the 18th International Conference on Electronic Business* (pp. 106–114). ICEB, Guilin, China, December 2–6.

Johnson, R. D., Stone, D. L. & Lukaszewski, K. M. (2020). The Benefits of eHRM and AI for Talent Acquisition. *Journal of Tourism Futures.* Retrieved from www.emerald.com/insight/content/doi/10.1108/JTF-02-2020-0013/full/pdf (Accessed 13 April 2021).

Kaplan, A. & Haenlein, M. (2019). Siri, Siri, in My Hand: Who's the Fairest in the Land? On the Interpretations, Illustrations, and Implications of Artificial Intelligence. *Business Horizons*, 62 (1), 15–25.

Kapoor, B. (2010). Business Intelligence and its Use for Human Resource Management. *The Journal of Human Resource and Adult Learning*, 6 (2), 21–30.

Lee, K.-F. (2019). *Artificial Intelligence and the Future of Work: A Chinese Perspective, Data, Ideas and Proposals on Digital Economy and the World of Work.* Retrieved from www.bbvaopenmind.com/wp-content/uploads/2020/02/BBVA-OpenMind-Kai-Fu-Lee-Artificial-intelligence-and-future-of-work-chinese-perspective.pdf (Accessed 30th March 2021).

Marr, B. (2017). *The Future of Performance Management: How AI and Big Data Cobat Workplace Bias*. Retrieved from www.forbes.com/sites/bernardmarr/2017/01/17/the-future-of-performance-management-how-ai-and-big-data-combat-workplace-bias/'?sh=4a87d2c64a0d (Accessed 23rd March 2021).

Meister, J. (2019). *Ten HR Trends in the Age of Artificial Intelligence*. Retrieved from Ten HR Trends in the Age of Artificial Intelligence (forbes.com) (Accessed 30th March 2021).

Merlin, P. R. & Jayam, R. (2018). Artificial Intelligence in Human Resource Management. *International Journal of Pure and Applied Mathematics*, 119 (17), 1891–1895.

Meyer, D. (2018). *Amazon Reportedly Killed an AI Recruitment System Because It Couldn't Stop the Tool from Discriminating Against Women*. Retrieved from https://fortune.com/2018/10/10/amazon-ai-recruitment-bias-women-sexist (Accessed 23rd March 2021).

Palmu, E. (2020). *Human Resource Management with Artificial Intelligence*. Retrieved from www.theseus.fi/handle/10024/338353 (Accessed 20th March 2021).

Pillai, R. & Sivathanu, B. (2020). Adoption of Artificial Intelligence (AI) for Talent Acquisition in IT, IITes Organisations. *Benchmarking: An International Journal*, 27 (9).

Prakash, C., Saini, R. & Sharma, R. (2021). Role of Internet of Things (IoT) in Sustaining Disruptive Businesses. In R. Sharma, R. Saini & C. Prakash, V. Prashad (Eds.), *Role of Internet of Things (IoT) in Sustaining Disruptive Businesses* (1st ed.). New York: Nova Science Publishers.

Prentice, C., Dominique Lopes, S. & Wang, X. (2020). Emotional Intelligence or Artificial Intelligence – An Employee Perspective. *Journal of Hospitality Marketing & Management*, 29 (4), 377–403.

Pribanic, E. (2018). *Future of AI in Corporate Training and Development*. Retrieved from www.techfunnel.com/hr-tech/future-of-ai-in-corporate-training-and-development/ (Accessed 23rd March 2021).

Rana, G. & Goel, A. K. (2015). Birla Milks Its Employees for the Creamiest Leaders of the Future: Internal Talent Has a Vital Role in Company Development. *Human Resource Management International Digest*, 23 (2), 9–11. https://doi.org/10.1108/HRMID-01-2015-0002.

Rana, G. & Sharma, R. (2019). Emerging Human Resource Management Practices in Industry 4.0. *Strategic HR Review*, 18 (4), 176–181.

Rao, V. S. P. (2013). Human Resource Management. *Excel Books*, 37.

Sharma, R., Singh, S. & Rana, G. (2019). Employer Branding Analytics and Retention Strategies for Sustainable Growth of Organizations. In *Understanding the Role of Business Analytics*, pp. 189–205. Singapore: Springer. https://doi.org/10.1007/978-981-13-1334-9_10

Tambe, P., Cappelli, P. & Yakubovich, V. (2018). Artificial Intelligence in Human Resource Management: Challenges and a Path Forward. *SSRN Electronic Journal*. Retrieved from https://papers.ssrn.com/sol3/papers.cfm?abstract_id=3263878 (Accessed 23rd March 2021).

Tzafrir, S. S. (2007). The Relationship Between Trust, HRM Practices and Firm's Performance. *The International Journal of Human Resource Management*, 16 (9), 1600–1622.

Vrontis, D., Chrostofi, M., Pereira, V., Tarba, S., Makrides, A. & Trichina, E. (2021). Artificial Intelligence, Robotics, Advanced Technology and Human Resource Management a Systematic Review. *The International Journal of Human Resource Management*, 1–30. Retrieved from https://www.researchgate.net/publication/349383945_Artificial_intelligence_robotics_advanced_technologies_and_human_resource_management_a_systematic_review

Wang, Y., & Kosinski, M. (2018). Deep Neural Networks Are More Accurate than Humans at Detecting Sexual Orientation from Facial Images. *Journal of Personality and Social Psychology*, 114 (2), 246–257.

Yano, K. (2017). How Artificial Intelligence Will Change HR. *People & strategy*, 40 (3), 43. Retrieved from www.cfsearch.com/wp-content/uploads/2019/10/James-Wright-The-impact-of-artificial-intelligence-within-the-recruitment-industry-Defining-a-new-way-of-recruiting.pdf (Accessed 21st March 2021).

Yawalkar, V. V. (2019). A Study of Artificial Intelligence and Its Role in Human Resource Management. *International Journal of Research and Analytical Reviews*, 6 (1), 20–24.

Zielinski, D. (2017). *Get Intelligent on AI: Artificial Intelligence Can Boost HR Analytics But Know What You're Buying.* Retrieved from www.shrm.org/hr-today/news/hr-magazine/1117/pages/artificial-intelligence-can-boost-hr-analytics-but-buyer-beware.aspx (Accessed 22nd March 2021).

9 AI and Business Sustainability
Reinventing Business Processes

D. Krishnaveni, V. Harish, and A. Mansurali

CONTENTS

9.1 Introduction 153
9.2 Review of Literature 154
    9.2.1 Artificial Intelligence 154
    9.2.2 AI and Business Sustainability 155
9.3 Objectives 156
9.4 Achieving Sustainability 156
9.5 Discussion 159
    9.5.1 Smart Mining 159
    9.5.2 Futuristic Agriculture 160
    9.5.3 Transforming Construction 162
    9.5.4 Revolutionizing Manufacturing 163
    9.5.5 Strategic Retailing 164
    9.5.6 Revamping Media and Entertainment 166
    9.5.7 Remodeling Financial Services 167
    9.5.8 Reshaping Education 168
9.6 Adverse Impacts of AI on Sustainability 169
9.7 Conclusion 170
References 170

9.1 INTRODUCTION

The Fourth Industrial Revolution is promising to change the way we do things on a scale and at a speed that we have not heretofore witnessed. Neither industry nor the academic world is clear on what the transformation will result in, nor who all will be impacted and at what levels. However, disruptions are being seen in several sectors, and the breadth and depth of these transitions is mind-boggling! In all this uncertainty, the one demonstrable trend is the thinning of the boundaries between what was clearly thought of before as being physical, digital or biological.

With substantial increases in computing power, storage capacity and easy access to information, the last few years have seen the rise of terms such as "artificial

DOI: 10.1201/9781003145011-9

intelligence," "quantum computing," "smart computing," "Internet of Things," and many more. Such breakthrough technologies are the vehicles that are moving us toward realizing possibilities that we would have normally relegated to the worlds of fiction and fantasy. These changes have accelerated the growth and profitability of sectors like banking, healthcare, retail, entertainment, telecom and travel, to name a few. It has now become easier to engage with customers and employees anywhere at any time of the day, to set up smart supply chains that are quick to respond to changes, to redesign business process workflows that are efficient and effective, and to build agile businesses that can adapt based on business intelligence and insights. The focus has shifted to empowering employees with an integrated outlook that can help them decide how to lower expenses while not compromising on the quality and quantity of output.

The fundamental aspect of the Fourth Industrial Revolution that is bringing about automation and causing large-scale changes is centered around data—its creation, documentation and transmittal. The infrastructural developments in the last few centuries have laid a slow but reliable foundation, that hand in hand with the advances in scientific know-how, have led to an inflection point in the technological evolution of the planet. The real impact however, is not just in the manner in which businesses are run or the profits that they can generate—it goes beyond commercial interests to touch upon the quality of life of people, many of whom do not even comprehend the myriad ways in which they are being impacted. But technology—specifically, AI—has the ability to make the lives of people all over the world more comfortable, particularly with its ability to analyze and predict change, thereby reducing uncertainty and fear.

Forecasts are suggesting that the increased use of connected devices, more customized use of IoT and shrinking internet access costs are driving the Fourth Industrial Revolution forward, with some estimates placing the Industry 4.0 market size at around $267 billion in 2026 (Mordor Intelligence, 2020). The COVID-19 pandemic has accelerated the pace at which automation and technological dependence have been accepted by industries as they strive to become more agile and resilient to environmental/societal disturbances and market shocks.

9.2 REVIEW OF LITERATURE

9.2.1 Artificial Intelligence

The Fourth Industrial Revolution is slated to disrupt industry trends with heavy reliance on newer advances in artificial intelligence (especially machine learning and deep learning); IoT (especially industrial IoT or iIoT); cloud computing and storage; virtual, augmented and mixed reality; cognitive computing; simulations; autonomous robots; Big Data; cyber-security; and system integration (i-scoop, 2021).

Broadly speaking, artificial intelligence is the ability of a computer or a machine to simulate human functioning—the machines are constantly learning from their experiences, comprehending and responding to varied inputs, and making decisions the way a human being would be expected to (TechGig, 2020). It is a set of algorithms that helps machines perform human tasks such as solving problems, even

AI and Business Sustainability

with large amounts of unstructured data. AI is already playing a big role in our lives, and while it will render some job roles obsolete, it is also set to create many high-value opportunities for those who are knowledgeable. AI can help transcend the barriers of language, time and geographical distance to deliver more creative and innovative solutions for smarter businesses.

Machine learning is that part of AI that deals with discovering patterns in huge amounts of data (Faggella, 2020). Using statistical processing, algorithms help machines discover features, patterns, trends or sequences in different types of data—text, numbers, images or even videos. Pinterest, Facebook, Twitter and Google are some of the famous organizations that have used machine learning in various capacities. It has been extensively employed in image recognition, speech recognition, medical diagnosis, buying and selling of financial assets to leverage price differences in different markets, and in many more areas where analyses and predictions are required to be made on the basis of a multitude of variables (Guerrouj, 2020).

One subset of machine learning that has people sitting up and taking notice across different sectors is deep learning. Deep learning is based on the way the human brain's neural networks function (Goodfellow et al., 2016). This comes very close to imitating human outputs in operations such as predictions and classifications. Automated translations into multiple languages and autonomous or self-driving vehicles are examples of deep learning. Other real-world applications of the technology include computer vision in gaming, news aggregation based on user preferences, language identification and analysis, and bot training (Vazirani, 2018).

There are several ways in which AI and its subsets are transforming human life as part of Industry 4.0. While there are definitely going to be impacts on the financial aspects of businesses because of AI, of more pressing concern are the probable implications on the other sustainability aspects such as the environmental as well as the societal pillars.

9.2.2 AI AND BUSINESS SUSTAINABILITY

Going by the United Nations' Brundtland Report released in 1987 (Britannica, 2021), sustainable development refers to the fulfilling of the needs of the current generation without jeopardizing the ability of future generations to fulfill their needs. This is based on the three pillars of environmental, economic and social equity (Purvis et al., 2019). Environmental stewardship refers to responsible utilization of the Earth's resources. The emphasis here is on creating a circular economy whereby the output of one process is used as the input for another, without any generation of waste. Economic sustainability deals with the generation of profits in such a way that environmental conservation and social justice is maintained. Social equity refers to the care and concern for welfare of the individual, as well as the society as a whole.

Business sustainability is a general term used to denote the long-term actions and strategies that organizations are focusing on that go beyond mere generation of profits to wanting to create a meaningful impact on society that will help them stay relevant for a long time (Bansal & DesJardine, 2014; Sharma et al., 2019). It is now being looked at as a source of competitive advantage, one that will help organizations attract new investors, retain their employees and also build a good corporate reputation. A report by management consulting firm McKinsey (McKinsey & Company, 2011) suggested that the motivations for business sustainability were not limited to having a good reputation in the marketplace, but also in order to generate new growth opportunities and gain operational efficiencies.

Artificial intelligence has a big role to play in the achievement of business sustainability. It has the potential to help in the supply of clean water, food and high-quality health services to people in different parts of the world. It can aid in the planning of smart cities that can help conserve energy and also shift to renewable sources and circular economies. The United Nations' Sustainable Development Goals—such as quality education, affordable and clean energy, economic growth, innovation and infrastructure, climate action and many more (Goralski & Keong Tan, 2020)—will become easier to attain by harnessing the power of AI. Specifically, in business, AI can be used to design low-carbon emission systems in different aspects of production, distribution and consumption.

9.3 OBJECTIVES

While several businesses have incorporated the triple bottom line approach in their work, there are many which have yet to do so. Given that sustainability-oriented businesses can lead to more satisfied customers and employees, and also to higher financial returns, this chapter is going to delve into the contribution of AI in different economic sectors and industries to understand the reach and scope of AI in sustainability. Specifically, the objectives of this chapter are the following.

To understand the current challenges that are acting as impediments to business sustainability in different economic sectors.
To analyze the role of AI in helping businesses become more sustainable.

9.4 ACHIEVING SUSTAINABILITY

The three-pillar model of sustainability suggests that if the social, economic and environmental pillars are represented as three intersecting circles, the area of intersection of all three circles is where we will find sustainability. However, modern businesses often find that they fall under one of the contexts shown in Figure 9.1. Sometimes the emphasis is often on one of the three pillars to the exclusion of the other two, or vice versa.

Organizations have understood that sustainability is now a business imperative and that achieving it can reap many advantages such as higher efficiencies, retention and attracting talent, greater branding and overall better stakeholder relations (Rana et al., 2021). All businesses are now faced with environmental challenges such as uncertainty about the future in the form of unpredictable market trends and changing regulatory and compliance requirements. Internally, they are struggling to contain the right competencies to survive in an agile environment while providing customized and high-quality services to customers. The business environment is also crowded with data, technological developments which need to be tracked, and

AI and Business Sustainability

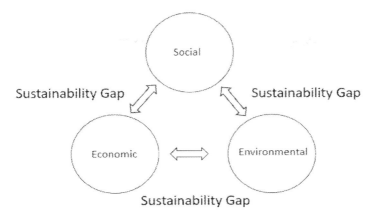

FIGURE 9.1A Lack of sustainability in social, economic and environmental aspects.

FIGURE 9.1B Either social and environmental sustainability or economic sustainability, but not both.

FIGURE 9.1 Sustainability gaps in modern businesses.

FIGURE 9.1C Either social and economic sustainability or environmental sustainability, but not both.

FIGURE 9.1D Either environmental and economic sustainability or social sustainability, but not both.

FIGURE 9.1 (Continued)

AI and Business Sustainability

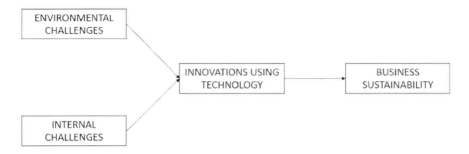

FIGURE 9.2 Possible pathway to business sustainability.

intense competition. Given these challenges, many businesses are turning to technology for achieving sustainability (Figure 9.2).

9.5 DISCUSSION

Eight economic sectors spanning mining, agriculture, manufacturing, entertainment, and so on, have been identified for the following analysis of the challenges they are faced with, and how AI can help with attainment of business sustainability.

9.5.1 SMART MINING

Mining, the process of extracting minerals, metals and other geological matter, is a primary economic activity that acts as a source of raw materials for several other industries. The major challenges facing the sector today include fickle market prices and resultant fluctuations in profitability; the decreased attractiveness of the sector as an employment sink when compared to other professions; the maturation of mines, which has led to circumstances that higher quality ore is less readily available; and intellectual property and other competitive concerns that are impeding innovation (Deloitte, 2020). Mining corporations are struggling to run viable operations that can keep up productivity while facing flagging returns. In the face of such problems, AI has the potential to help in various ways.

Many decisions in the course of mining are taken on the basis of data collected through human inspection of sites, which is a time-consuming process. The use of strategically placed sensors can gather real-time data that can facilitate a quicker and more accurate decision-making process that will also be consistent, and many times also proactive. With increased certainty, mineworkers need to take fewer risks with their health and safety. The additional advantage with AI is that similarly to humans, machine learning algorithms can be trained, additionally augmenting efficiency.

Among the different processes involved in the mining process, the most energy intensive is that of ventilation. By identifying trends in past and current data, energy peaks can be predicted, making it a much more energy efficient process. Such AI technology is helping to greatly reduce the carbon footprint of the mining sector as a whole. Furthermore, the constant capturing of environmental and structural data through sensors and other equipment will build on the body on existing knowledge to generate newer and better insights that can consequently enhance optimized use of resources.

RockMass Technologies, Ionic Engineering, Thorough Tec, Shift Inc. and Wipware are some of the organizations that have implemented AI technologies in their day-to-day mining operations. Freeport-McMoRan is a U.S.-based copper mining organization that has always relied on extensive data capturing and analysis. With the use of sensors and other equipment on many types of stationary and moving vehicles and machinery, real-time data was recorded over time. Manual analysis of the data would have been time consuming, expensive and quite unlikely to detect trends that may have been deeply entrenched. This is when the organization decided to use the help of McKinsey data analysis to and suggest changes that would improve productivity and profitability in the context of falling copper prices in the market. Going by the recommendations suggested by data modeling algorithms, the organization successfully increased mineral production to rates that were never achieved before. When combined with agile operating principles, the organization was able to utilize machine learning outputs to create functional solutions swiftly and efficiently (McKinsey & Company, 2021).

In every stage of mining, be it prospecting, reaching or breaking the ore, bringing the ore up to the surface or dressing and smelting the ore, AI is helping organizations to automate processes. By connecting data retrieved from several sources such as satellite images and aerial pictures, machine learning and cognitive computing is used to design optimized solutions and techniques to increase reliability, accuracy and safety of operations.

9.5.2 Futuristic Agriculture

Even though agriculture is one of the oldest lines of work, changes in the way it is practiced occurred very gradually for the first few millennia. With a burgeoning global population that is expected to touch nine billion by the year 2050, there is high pressure on the sector to improve productivity and satisfy demand for food as well as industrial raw materials, while using fewer inputs and resources.

As a sector, agriculture is plagued by uncertainties on multiple fronts. Farmers have to track a multitude of variables such as sunlight, heat and humidity; soil moisture, nutrient content and pH; rainfall levels; wind speed and direction; animal, bird and insect populations and behavior; market rates and subsidies; changing consumption patterns of output; cost of inputs like fertilizers, insecticides, weedicides; and so on. For one farmer to keep tabs on field, weather and market conditions can be very overwhelming. This is where AI is stepping in. In fact, a whole new type of farming called precision agriculture—which is focused on exact and specific inputs at different stages of plant growth—has come into existence (Schmaltz, 2017).

Precision agriculture is especially suitable for those land holdings that fall under the marginal, small and medium categories, as it is nearly impossible to achieve economies of scale in these cases. This system is essentially classified as low input and high efficiency, which are the founding pillars of sustainability.

AI-based surveillance systems have been set up in farms to identify animal or human intrusions or breaches. This form of video vigilance, which can further be triggered to drive remedial measures, is necessary to protect crops from predators and theft. Through an elaborate set of sensors placed in the ground and also through the use of drones that perform aerial surveillance, it is now possible to capture data about factors like soil moisture, mineral content, pH levels, and so on, that aid in understanding crop growth patterns and forecasting crop output levels. In fact, it has now become possible to determine to a large extent just how much yield can be obtained from a piece of land for a particular crop based on current soil conditions and weather conditions (Forbes, 2021). Solar pumps that have a low ecological footprint can be prompted to work based on knowledge of humidity and moisture levels (Mungerwal & Mehta, 2019) resulting in less irrigation water consumption.

In general, a substantial percentage of crop yield is destroyed by pests and diseases every year. Having prior knowledge about pest and disease movement can aid farmers reduce crop losses and also get access to timely expert advice. AI is now being used to detect such adverse events through the use of images and also to warn farmers of likely attacks. Smart pest monitoring systems have been designed that can count the number of pests, identify their types and location in the field, and then send out alerts to farmers with this information. Plantix is another app that uses data provided by the International Crops Research Institute for the Semi-Arid Tropics (ICRISAT), which has helped around 75,000 farmers in India gain access to precise information about pests and diseases (ICRISAT, 2020).

Climate is yet another factor that has a major impact on agriculture. With global warming, many farmers have been faced with issues like increased unpredictability in weather, unexpected changes in climate patterns, sudden and heavy downpours and other climatic deviations. By analyzing past trends in rainfall, temperature, wind, air pressure and humidity, and also by monitoring the current patterns through the use of sensors, farmers will have access to more accurate and timely information which can help them determine the appropriate times for planting, adding fertilizer, harvesting, and so on. Farmers can also take advantage of certain conditions such as wind direction and intensity for fertilizer application.

By using data generated by sensors and global positioning systems, it is possible to determine which chemicals, fertilizers and herbicides, and also how much water, needs to be provided to which parts of the farm. This technology, known as variable rate application, is an integral part of precision agriculture and provides invaluable information to the farmer, besides helping him save on costs of input (Igor, 2018).

Agribots are smart devices that use image processing to differentiate between crops and weeds, and which will then spray weedicides selectively. Such selective spraying is a good farming practice that will help bring down resistance to weedicides and herbicides in the future. Harvesting is yet another area in farming which is becoming increasingly expensive, with many people not preferring to work in this sector. In order to overcome this shortcoming, agribots have been designed that can pick fruits and vegetables, overcoming the restrictions of fatigue, cost and inexperience. Similarly, it is now possible to run machines like tractors and other heavy machinery on the fields by operating them remotely, ensuring that work goes on without hindrance and reliance on external labor.

Finally, while picking and during the sorting process, computer imaging can be used to analyze the size, shape and color of crop output to help in categorizing it into different grades which will fetch different rates at the market. Besides this, AI is used in irrigation management, often bringing down water consumption to a fraction of what was being used prior to using the technology.

However, while AI has been known to bring to the fore several advantages for farmers, there are downsides such as high costs of setting up and maintaining equipment, learning to use the equipment and making the right inferences from the wealth of data that will be generated and scalability.

9.5.3 Transforming Construction

Among the biggest challenges faced by the construction sector today are issues like shortage of skilled labor, increasing cost of input materials, safety-related issues and decreasing profits arising from lowered productivity (Construction Placements, 2019). Additionally, there are a host of other problems such as delayed shipment of materials, weather deviations wrecking preplanned schedules and changing customer awareness about concepts such as building ratings on the basis of energy consumption or environmental impact that have forced the construction sector to step back and relook at what can be done to better serve customers.

Every construction activity is an exercise in project management. By assimilating AI in the project management function, it is possible to relegate mundane and low-value–addition tasks to machines, freeing up the manager's time for bigger and more strategic decisions. Given the ability of AI to understand and map competencies and skills of employees, it can be used to identify the most suitable candidate for performance of different jobs on the basis of their skill sets and their availability. With the use of drones and other surveillance equipment, it is possible for AI to capture data about job sites and perform geospatial analyses that can help construction workers anticipate issues and come up with workarounds earlier, in order to stay on schedule (Invonto, 2018). Also, in the event of unexpected delays, AI algorithms can help run "what if" scenarios and generate probable solutions that can be implemented to get back on track.

Newer advances in vehicle technologies have helped to semi-automate machines such as bulldozers which are now increasingly being used to execute monotonous and banal jobs such as bricklaying and pouring concrete. Autonomous machines and bots are also being used to produce partial construction off-site which will later be incorporated into the main structure, further freeing up workers for other jobs. AI, especially IoT, in the form of wearable sensors can help monitor different conditions and contribute to increasing worker safety at the site.

With the ability to predict more realistic timelines and expense estimates, cost overruns can be kept to a minimum with the use of AI. Advanced machine learning software products are now available that can integrate the plans of different teams such as electrical, mechanical, architectural, engineering and plumbing into 3D models to ensure that all plans are in sync. "Generative design" is the term given to a

AI and Business Sustainability   163

repetitive process whereby an AI program produces certain designs that are designed to fit the constraints given by the customer. Each one of these designs is a possible solution for the construction company. They can pick any of these designs or further refine them by fine tuning some of the constraints to produce even more customer-friendly solutions. Eventually, this type of technology might render architects and other related designers redundant.

Green construction or sustainable buildings are terms that are increasingly becoming popular in the construction sector. Customers today are more environmentally sensitive than ever before, and they often demand that building constructions be resource economical and environmentally accountable. The idea is not just to be environmentally responsible during the design and construction phase, but also during the maintenance, renovation and demolition stages. In order to make it easy for customers to know about the buildings, several rating agencies have come up with different certifications such as the Leadership in Energy and Environmental Design (LEED), the BREEAM (Building Research Establishment Environmental Assessment Method), EDGE (Excellence in Design for Greater Efficiencies) and Global Sustainability Assessment System (GSAS). The primary aim of all these certifications and rating agencies is to promote environmental stewardship by ensuring economical utilization of resources such as water and energy, to positively impact the health and safety of occupants and to decrease waste generation and pollution.

Currently, AI is not as extensively used in construction as it will be in the future. There is a dearth of sufficient "Big" Data and data analysts to train algorithms on different aspects of construction. However, smart construction technologies are slowly but surely making inroads into the sector as the potential for savings, as well as for better designed outputs, increases radically with the use of such technologies.

9.5.4 Revolutionizing Manufacturing

The last couple of decades have witnessed a sea change in manufacturing processes that have helped businesses radically change the way they operate. With globalization, the focus has moved toward cost-effective production without compromising on quality, which can help in fending off revenue volatility. There is also an increased emphasis on having shorter and shorter production times to enhance competitiveness. With more regulatory restrictions and increased monitoring through inspections, and also related litigation costs, businesses are looking to improve quality continuously. Additionally, the sector has to be able to adapt to sudden changes in market conditions, suddenly increasing or decreasing quantities of production and also being able to accommodate high levels of customization.

Defect detection is a huge part of the quality improvement process. The emphasis has now shifted from manual inspections to validation by machines, which is more reliable, accurate, and consistent. Even the smallest of defects, which may often be missed by the human eye, can be captured in high resolution images and highlighted for corrective action. Such techniques are helping companies achieve their Quality 4.0 (The switchover to digital technologies to attain organizational excellence and quality goals.) targets while simultaneously reducing time-to-market.

A new concept called "digital twinning" has come up in manufacturing. This refers to a virtual representation of an actual product, often developed before the product, to understand its features better. The digital twin is often tweaked to analyze its impact on performance in order to determine the best possible version of the product. It permits customizations and prediction of performance without much cost. As in construction, "generative design" is also applicable in the area of manufacturing (CIO, 2018).

Predictive maintenance is yet another area where manufacturers are looking to incorporate AI. A large number of disruptions in manufacturing occur due to accidents or downtime, which can be largely forecasted by machine learning algorithms. Being able to predict the precise times when maintenance needs to be performed can help organizations save costs on unnecessarily frequent maintenance exercises, while at the same time, not delaying it for so long as to cause disruptions due to unwarranted downtimes.

Another important aspect of manufacturing is being able to forecast the quality and quantity of output to be produced within a time frame. By analyzing the quantity and quality of input material, AI algorithms can predict the quality and yield of output under varying conditions and constraints. It can also be used to understand what is the best way to reduce generation of waste and how to use resources optimally—in effect, making processes more efficient and further reducing costs.

Slowly, in keeping with the trend sparked by changes related to Industry 4.0, many businesses are utilizing increased automation of production processes. This has also helped in increasing the efficiency, reliability, accuracy and safety of these processes. With more automation, businesses will find it easier to be agile and make real-time adjustments in production requirements, not just in terms of production quantities but also with respect to customizations that can give organizations a competitive edge.

Outside of the manufacturing process itself, AI can help in streamlining the supply chain process, aiding in the identification and dispatch of timely bills of materials without human interference. By being able to analyze past trends in production, keeping in mind seasonality and other factors, AI can help predict demand patterns and advise on changes that have to be made accordingly for seamless production. By harnessing real-time trends and market conditions, AI can also assist in creating dynamic pricing structures that can help take advantage of supply shortages, thereby increasing profit potential.

Another very important aspect that is often not given the importance it deserves is customer analytics. Machine learning can be used to analyze customer consumption patterns, enabling the understanding of their purchasing behavior and further guarding against loss of any potential sales.

9.5.5 STRATEGIC RETAILING

Retailing is the process through which merchandise or some services reach the end consumer. While a few retailers produce the products they sell, most of them buy their goods from a manufacturer or a wholesaler and sell in small quantities. It has always been a competitive sector, with retailers vying with each other to achieve

AI and Business Sustainability

high sales volumes. Today's customers are spoiled by choice and are able to obtain timely and precise information about products and services before making purchase decisions. With brand loyalty becoming passé, retailers are left wondering about how to retain existing customers and also how to attract new ones.

Even though shopping was already well on its way to becoming accepted digitally (even in rural areas), the COVID-19 pandemic has accelerated the process of making e-retail more universally entrenched in the daily routine of consumers. Therefore, the trend is now to perform some part of purchase in the traditional brick-and-mortar stores and the rest through online means. Given the proliferation of multichannel buying options, customers are looking to have a seamless experience in purchasing while switching between channels (eTail, 2021).

Increasingly, retailers have had to go beyond customer delight to creating buying experiences that customers would want to repeat. Advertising costs shave also shot up, with marketing to be done across multiple media to reach all segments of the target population. A fine lines has to be drawn between overwhelming customers with communication and piquing their curiosity to check out the product or service. With so many channels and technologies on the market, integrating them together and ensuring that the communication from the business is uniform across the spectrum is a challenge unto itself.

Nowadays, many customer relationship management (CRM) software products use AI to study consumer behavior and determine when they are likely to purchase different products. Retailers are looking at using "edge computing" whereby computation and processing is done physically close to where the data is generated to capitalize on the twin advantages of faster response times and to reduce bandwidth use. With the combination of AI-driven CRM and edge computing, retailers can help deliver great buying experiences for customers, while also gaining insights into where they can innovate and create more opportunities for sales. This can also help to personalize the experience, making the customer want to come back to the store.

Retailers can also look at cutting down on operational expenses by replacing certain manual processes, such as billing, which can be automated. Several businesses are already using AI-powered chatbots to make suggestions about related accessories or higher versions of merchandise to customers. When it comes to inventory management, AI has multiple roles to play. Using machine learning algorithms, it is possible to forecast demand for merchandise and raise inventory levels accordingly. AI can also identify real-time patterns or trends in sales and make necessary inventory stocking corrections.

Due to high competition levels between businesses in retail, activities such as promotions, discounts, sales, and so on, are commonplace in the sector. By using AI, prices of different goods can be optimized to increase their sales, while at the same time improving profitability of operations. This can also help to cut down losses due to unsold stock or situations where the customers' requirements were not available in store for sale, thereby streamlining supply chain management. Machine learning can also help identify which products are usually bought in combination with others and help within-store stocking arrangements. While price optimization usually pertains to the current stock levels and promotional offers, AI can also help to forecast prices

166 Reinventing Processes Through AI

for the future based on predictive analytics, helping customers decide when they want to purchase the product.

For customers, AI can help in finding the location of items or items similar to what they have in mind by uploading text, images or voice instructions (SPD Group, 2021). In the case of clothing, where customers generally like to try on a garment before purchase, AI can help in scanning their body dimensions and virtually showing the fit of the garment to aid in purchase decisions. Businesses like Walmart have also been recently using AI to read facial cues of customers through cameras in order to identify their moods and understand their satisfaction levels with the entire shopping experience.

9.5.6 Revamping Media and Entertainment

The media and entertainment sector has always been at the forefront of technological experiences. However, faced with challenges such as piracy and widely available free content, businesses in this sector are now forced to re-evaluate their business model mix and walk the fine line between subscription-based, one-time payment–based and free content generation models. With video on demand and over-the-top media garnering more subscribers, the traditional television industry has to rework both its content on offer and its advertising policies to stay profitable.

As people have now become accustomed to breathtaking and sensational experiences on screen, the minimum acceptable threshold of acceptance of entertainment content has also gone up. Not only should the content be of high quality, but it also needs to be personalized to a certain extent to keep viewers engaged. The onus is on the industry to provide continuous generation of high quality, creative content that has not been anticipated by audiences. In the case of news media, viewers are now expecting round-the-clock credible and unbiased coverage. With multiple channels throwing a wide variety of content at viewers, it is increasingly difficult for media and entertainment houses to hold on to the attention and loyalty of viewers (Chazen, 2020).

In the movie industry, AI is already being used to generate scripts and synopses, and identify character names that are likely to be well liked among the target audience. On the other hand, while being fed a script, machine learning can identify discrepancies in the storyline and suggest appropriate dialogue. Based on situations described in the script, AI can make location and set recommendations—and also compare attributes of the characters with the skill sets of available actors, and suggest matches. All of this will greatly simplify the preproduction stage in filming. Further potential AI uses are in understanding popular music scores and what really resonates with the audience in different contexts, and also in editing. Another huge advantage of AI in movies would be in analyzing the audiences in different places and designing promotional campaigns that will improve the chances of the movie becoming a hit.

The biggest contribution of AI to the sector, however, could be in the understanding of consumers, their preferences and interests, and the ability to gauge their moods and suggest content accordingly. Being able to predict what the audience is likely to want to watch, and through which medium or devices, can help in packaging suitable

AI and Business Sustainability

content—with some subtle and unobtrusive advertising thrown in the mix. AI can also help to track what kind of programs consumers are interested in, when they lose interest, if they have shifted to other channels—and if so, why. Understanding customers is key to design the targeted advertising campaigns that are substantial revenue generators for such businesses.

Augmented reality (AR) is a technology whereby the real-world environment is supplemented with virtual objects such as images, sounds, and so on. Originally developed for the gaming industry, AR is nowadays being used in multiple fields. Similar to AR is virtual reality (VR), a technology whereby a user is completely immersed in a digital environment. VR and AR, together known as mixed reality (MR), are now being used to train people through simulations in fields like medicine, warfare, engineering, and many others. Training through such simulations helps to prepare doctors, engineers, soldiers and other professionals without exposing them to the real dangers in their respective fields, and helps in cutting down on the time and expense involved in recreating critical situations for practice (Sharma & Rana, 2021; Franklin Institute, 2021).

Specifically, in the gaming industry, viewers are expecting more and more realistic experiences, in order to retain their attention in the game. Three-dimensional visualizations are testing the capabilities of game developers who are always looking to see how they can deliver the unexpected to the user. AI-powered algorithms are now designed to take in inputs not just through the console or other physical activity, but also through voice commands (AIThority, 2020). Companies like Latitude are in the process of developing "open-ended" games that are heavily dependent on machine learning to determine how to proceed. Another aspect to be kept in mind is the fact that unlike in the past when gaming was primarily done through computers or consoles, nowadays a variety of other gadgets such as mobile phones and tablets are being used for gaming. Therefore, users want to be able to experience a seamless transition from one medium to another.

9.5.7 REMODELING FINANCIAL SERVICES

The financial sector, which was largely the domain of banks in the past, is now crowded with fintech and non-banking companies. With increasing competition, there is a change in the mindset to move away from traditional or legacy systems to using digital systems that can speed up processes which are also secure and hassle-free for the user (Burlakov, 2021). Institutions are competing with each other to design and offer promotions calculated at retaining and attracting new customers and shareholders.

The profile of the typical customer in this sector is also changing. Younger customers are more tech-savvy and prefer to perform many of their transactions online, without the necessity of visiting physical offices. The biggest threat, however, is the question of security of online transactions. Cyber-crime is on the rise, with hackers and criminals being able to break into the most secure of systems and costing financial institutions a lot of money. Machine learning algorithms are increasingly being used to study user behavior and identify suspicious activity that is then flagged as potentially a fraudulent transaction. A second type of fraud is money laundering—a crime that was relatively hard to detect before AI was used. With AI, it is easier to identify suspicious activity and reduce investigative time (Towards Data Science, 2019).

Due to computational restrictions in the past, risk analyses were often done with the assumption that markets behave linearly. However, with AI, such computations can be performed in a real-world, non-linear context to provide much better analysis. When it comes to social media and other data gathered from the web, financial institutions can utilize AI to filter out repetitive and fake news, in order to understand patterns and trends among users' opinions, based on which future company policies will be designed.

AI can be used in the evaluation of credit worthiness of users based on their past records and their current digital footprint, such as phone records. Credit services and interest rates can be customized to correspond to the user's habits. This has huge implications in loan disbursement and underwriting, where financial institutions accept financial risk for a fee. Another area where the power of AI can be leveraged is in trading. AI can produce data-driven, timely and low-risk forecasts while considering a whole host of variables, something that would not be possible manually. Trading portfolios can be designed faster and corrective actions taken in real time to reduce or prevent losses. A new term, "bionic advisory," which refers to the combination of in-depth machine calculations tempered by insights derived from human experience, is becoming quite popular. It is promising to disrupt the financial services sector by bringing to the fore advantages that neither machine-based nor human-based inputs can individually account for.

In their effort to be more accessible to their customers, banking services today are using chatbots to help customers with their queries and issues. There are many apps that have been designed—sometimes by financial institutions—to help users manage personal finances, aiding in the creation of financial plans, setting milestones and reminders, and tracking expenses. Finally, thousands of hours' worth of mundane, repetitive work can be automated, and employees engaged in such processes can be diverted to other important work, cutting down on operational costs at all financial institutions.

9.5.8 RESHAPING EDUCATION

The education sector has the onus of preparing students to become good employees, entrepreneurs or professionals in a variety of sectors. With expenses related to infrastructure and other resources going up, schools, colleges and other educational institutions are trying to measure if there has been a corresponding improvement in performance of their students to match the rise in costs.

Starting with the most basic of functions for teachers, which is evaluation of answer scripts or homework, AI can be trained to take over, leaving teachers and teaching assistants with more time to focus on making sessions more interesting and engaging (Teach Thought, 2014). By analyzing answer scripts and assignments, AI can identify which segments of the syllabus have been clearly understood and which segments need to be looked at again. It can be used in development, as well as dissemination, of course material and also for nudging students to read up about related

AI and Business Sustainability

events in current affairs. Another avenue for reducing the burden on teachers is to make AI perform administrative duties. Chatbots could be trained to act as counselors, mentors and guides that help students become better people.

Education can also be more personalized to keep up with the pace of learning of the student. With adaptive games and learning apps that can change the pace and spend more time on topics where the student is weak, AI-driven educational software can be a big boon to supplement teacher-taught material in class. In those cases when students are afraid of attempting answers or solutions because of fear of ridicule by peers or teachers, AI can be used to train the students. AI does not discriminate or embarrass, and can be made to explain the same topic in several different ways to ensure that the student understands. There is also some research going on in the creation of customized digital textbooks that use AI-powered evaluations of students in their design (Johnson, 2019).

In the case of higher education admissions, a large part of the process can be automated with AI evaluating admission tests and statement of purpose essays of applicants. Chatbots could also be trained to help students identify the appropriate career choices to make and the proper colleges to apply for admission.

9.6   ADVERSE IMPACTS OF AI ON SUSTAINABILITY

The preceding sections deal with how AI is making inroads into a variety of sectors such as mining; agriculture; construction; manufacturing; retail; media, gaming and entertainment; financial services; and education. In each of these sectors, AI is able to increase productivity, safety of workers, accuracy and timeliness of project completion; reduce inefficiencies; and also maintain consistency. However, not all of these benefits come without strings attached.

AI as a technology is expensive and not readily available or affordable in all parts of the world. Just as the internet created a digital divide, splitting the world into people who had access to a larger number of resources and those who did not, AI could become another differentiator that gives enormous computing potential and data-driven insights to those with the means to access it versus those who do not (Vinuesa et al., 2020). While the wealthier nations in the world are investing in AI heavily and will reap the benefits, the poorer nations will be left struggling, and the gap between the two groups will only widen further.

There will always be groups with vested interests who will use AI to further their own aims. It is possible to study people, analyze their weaknesses and influence or exploit them for the benefit of entities like governments and business organizations. People may not even be aware that their movements and activities are being monitored, or that they are being influenced to make certain choices, raising the ethical dilemma of just how intrusive AI can be. Further, machine learning algorithms will try to learn from existing data to come up with patterns and trends that it can apply to future data. The presence of biases and prejudices—which will be apparent in past data—could very well be projected on future data, while the opinions of users may have changed in the meantime. For AI algorithms to understand the principles of diversity, social inclusion, affirmative action, gender equality, wage parity and other socially important factors will be a challenge. For example, by analyzing the social media footprint of a person, AI algorithms may have identified them as having certain tendencies or possessing certain traits. Further suggested reading and targeted advertising might be so specifically aimed at the person that they soon start interacting only with the sites that already have similar biases and traits. This robs the person of the chance to ever change their opinions through serendipitously landing on a site that may have prompted them to question their heretofore rigidly held beliefs.

With increasing use of AI, the nature of jobs will change. Low-skilled, labor-intensive jobs will be lost, while high-skilled jobs will be created. Unfortunately, as education is still not universally accessible to all, many people will be rendered without the means to earn a living, further expanding economic inequality. Even for those with access to education, they will have to learn how to interpret AI outputs and design solutions on the basis of data-driven insights. With increased automation, the owners of businesses will benefit more than the employees of the business.

With respect to the environment, AI has the potential to study climate change patterns, analyze its causes and even recommend corrective action. By practicing precision agriculture, we could reduce farming inputs without compromising on the output quantity or yield quality. However, there is one major disadvantage of AI that does not often find a mention in literature. This is the huge carbon footprint in running AI. For every smart system that has to be developed and used by consumers, there are energy requirements, which when multiplied by the proposed number of users in the future, would become substantial. With the current rate of energy production from renewable energy sources, and the likely demand placed on the energy systems by AI-driven information and communication technology systems, it is quite possible that AI will be driven largely by non-renewable energy sources, which will nullify many of the environmentally sustainable achievements in the different sectors mentioned in this chapter.

9.7 CONCLUSION

AI has its share of believers and skeptics. However, it is a fact that it is already a part of our lives and will only play a bigger role in the future. Like electricity, the internet and many of the big innovations that turned out to be inflection points in the history of humankind, AI is also bringing about a paradigm shift in the way things are being done and perceived, impacting all aspects of life on earth. Given that it has the capability to soon outstrip humans, even in cognitive functions like decision-making, AI has to be carefully promoted to make life comfortable, better and thereby more sustainable for people.

REFERENCES

AIThority. (2020). *AIThority*. Retrieved from Understanding the Role of AI in Gaming: https://aithority.com/computer-games/understanding-the-role-of-ai-in-gaming/

Bansal, P., & DesJardine, M. R. (2014). Business sustainability: It is about time. *Strategic Organization*, 12(1), 70–78.

Britannica. (2021). *Brundtland Report*. Retrieved from Encyclopedia Britannica: www.britannica.com/topic/Brundtland-Report

AI and Business Sustainability

171

Burlakov, G. (2021). *Technorely*. Retrieved from 10 Challenges for the Financial Services Industry: https://technorely.com/financial-industry-challenges/

Chazen, D. (2020). *Verbit*. Retrieved from Solving Challenges in the Media Industry: https://verbit.ai/challenges-in-the-media-industry/

CIO. (2018). *CIO India*. Retrieved from 5 Ways Industrial AI is Revolutionizing Manufacturing: www.cio.com/article/3309058/5-ways-industrial-ai-is-revolutionizing-manufacturing.html

Construction Placements. (2019). *Construction Placements*. Retrieved from 5 Major Problems in the Construction Industry: www.constructionplacements.com/5-major-problems-in-the-construction-industry/

Deloitte. (2020). *Future of Mining with AI: Building the First Steps Towards an Insight Driven Organization*. Ontario, Canada: Deloitte.

eTail. (2021). *eTail Summit*. Retrieved from 5 Key Challenges Facing Retailers Today—And How to Solve Them: https://etaileast.wbresearch.com/blog/five-key-challenges-for-retailers-how-to-solve-them

Faggella, D. (2020). *Emerj*. Retrieved from What is Machine Learning?: https://emerj.com/ai-glossary-terms/what-is-machine-learning/

Forbes. (2021). *Forbes*. Retrieved from 10 Ways AI Has The Potential to Improve Agriculture in 2021: www.forbes.com/sites/louiscolumbus/2021/02/17/10-ways-ai-has-the-potential-to-improve-agriculture-in-2021/?sh=2d18fe9c7f3b

Franklin Institute. (2021). *Franklin Institute*. Retrieved from What's the Difference Between AR, VR, AND MR?: www.fi.edu/difference-between-ar-vr-and-mr

Goodfellow, I., Bengio, Y., & Courville, A. (2016). *Deep Learning*. Cambridge, MA: The MIT Press.

Goralski, M. A., & Keong Tan, T. (2020). Artificial Intelligence and Sustainable Development. *The International Journal of Management Education*, 18(1).

Guerrouj, L. (2020). *Machine Learning: 6 Real-World Examples*. Retrieved from Salesforce: Machine Learning: 6 Real-World Examples.

ICRISAT. (2020). *ICRISAT Happenings Newsletter*. Retrieved from Artificial Intelligence to Track Pests and Diseases in India: www.icrisat.org/artificial-intelligence-to-track-pests-and-diseases-in-india/

Igor, I. (2018). *Medium*. Retrieved from Variable Rate Application in Precision Agriculture: Variable Rate Application in Precision Agriculture.

Invonto. (2018). *Invonto*. Retrieved from Smart Construction: Build Smarter with Artificial Intelligence: www.invonto.com/insights/smart-construction-artificial-intelligence/

i-scoop. (2021). *Industry 4.0: The Fourth Industrial Revolution—Guide to Industrie 4.0*. Retrieved from i-scoop: www.i-scoop.eu/industry-4-0/

Johnson, A. (2019). *eLearning Industry*. Retrieved from 5 Ways AI Is Changing the Education Industry: https://elearningindustry.com/ai-is-changing-the-education-industry-5-ways

McKinsey &Company. (2011). *The Business of Sustainability*. Retrieved from McKinsey: www.mckinsey.com/business-functions/sustainability/our-insights/the-business-of-sustainability-mckinsey-global-survey-results#

McKinsey &Company. (2021). *McKinsey Insights*. Retrieved from Inside a Mining Company's AI Transformation: www.mckinsey.com/industries/metals-and-mining/how-we-help-clients/inside-a-mining-companys-ai-transformation

Mordor Intelligence. (2020). *Industry 4.0 Market—Growth, Trends, Covid-19 Impact, and Forecasts (2021–2026)*. Retrieved from Mordor Intelligence: www.mordorintelligence.com/industry-reports/industry-4-0-market

Mungerwal, A. K., & Mehta, S. (2019). *DownToEarth*. Retrieved from Why Farmers Today Need to Take up Precision Farming: www.downtoearth.org.in/blog/agriculture/why-farmers-today-need-to-take-up-precision-farming-64659

Purvis, B., Mao, Y., & Robinson, D. (2019). Three Pillars of Sustainability: In Search of Conceptual Origins. *Sustainability Science*, 14, 681–695.

Rana, G., Agarwal, S., & Sharma, R. (Eds.). (2021). *Employer Branding for Competitive Advantage: Models and Implementation Strategies* (1st ed.). Boca Raton: CRC Press. https://doi.org/10.1201/9781003127826

Schmaltz, R. (2017). *AgFunder*. Retrieved from What Is Precision Agriculture?: https://agfundernews.com/what-is-precision-agriculture.html

Sharma, R., & Rana, G. (2021). Revitalizing Talent Management Practices through Technology Integration in Industry 4.0. In R. Sharma, R. Saini, & C. Prakash (Eds.), *Role of Internet of Things (IoT) in Sustaining Disruptive Businesses* (1st ed.). New York: Nova Science Publishers.

Sharma, R., Singh, S., & Rana, G. (2019). Employer Branding Analytics and Retention Strategies for Sustainable Growth of Organizations. In *Understanding the Role of Business Analytics*, pp. 189–205. Singapore: Springer. https://doi.org/10.1007/978-981-13-1334-9_10

SPD Group. (2021). *SPD Stories*. Retrieved from Artificial Intelligence for Retail in 2021: 12 Real-World Use Cases: https://spd.group/artificial-intelligence/ai-for-retail/#1_Stores_can_become_cashier-free

Teach Thought. (2014). *Teach Thought*. Retrieved from 10 Roles for Artificial Intelligence in Education: www.teachthought.com/the-future-of-learning/10-roles-for-artificial-intelligence-in-education/

TechGig. (2020). *TechGig*. Retrieved from Understanding the Difference between AI, ML, and DL: https://content.techgig.com/understanding-the-difference-between-ai-ml-and-dl/articleshow/75493798.cms

Towards Data Science. (2019). *Towards Data Science*. Retrieved from the Growing Impact of AI in Financial Services: Six Examples: https://towardsdatascience.com/the-growing-impact-of-ai-in-financial-services-six-examples-da386c0301b2

Vazirani, J. (2018). *10 Real World Examples of Deep Learning Models & AI*. Retrieved from Futran Solutions: www.futransolutions.com/10-real-world-examples-of-deep-learning-models-ai/

Vinuesa, R., Azizpour, H., Leite, I., Balaam, M., Dignum, V., Domisch, S., & Nerini, F. F. (2020, January 13). The Role of Artificial Intelligence in Achieving the Sustainable Development Goals. *Nature Communications*, 11, 233.

Index

A

ability, motivation, opportunity (AMO), 10
administration, 71, 74
agriculture, 159–161, 169
AI-based systems, 20, 21, 23, 38
algorithm, 154, 155, 159–165, 167, 169, 170
artificial general intelligence (AGI), 2, 3, 21, 22
artificial intelligence (AI), 1, 2, 3, 8, 9, 20, 21,
  22, 23, 36, 40, 41, 42, 44, 53, 87, 95, 97, 98,
  99, 100, 101, 102, 103, 104, 105, 106, 107, 108,
  109, 119, 129, 132, 137, 138, 139, 140, 142,
  143, 145, 146, 147, 148, 154, 156; advantages
  of, 66, 67
artificial intelligence human resource
  management (AIHRM), 140, 143
artificial narrow intelligence (ANI), 2, 3, 21
artificial super intelligence (ASI), 2, 3, 21, 22
applicant tracking system (ATS), 145
augmented reality, 98
automation, 67, 69, 70, 71, 104, 105, 106, 107,
  108, 109

B

back propagation (BP), 145
backtracking search, 86
Big Data, 20, 34
Big Data analysis, 99
BigMart, 81
Bitcoin, 24, 27, 29, 32
blockchain, 20, 24, 25, 26, 27, 28, 29, 30, 31, 32,
  33, 34, 36, 37, 38, 100
blocks, 25, 26, 27, 29
Bluetooth, 46, 47, 48, 49
brain-computer interfacing (BCI), 42, 44, 45, 46,
  47, 48, 49
brain waves, 47
breakdown maintenance, 128, 129
business sustainability, 155, 156, 159

C

classification models, 87
classifiers, 54, 55, 56, 58, 59
cloud, 23
cloud computing, 37, 98, 99, 100, 105, 176
competitive advantage, 108, 109, 110, 155
computer science, 40, 42
computer vision, 2, 3
computing, 153, 154, 160, 165, 169

consensus, 24, 25, 27, 29, 30, 33, 34, 36
construction, 162–164, 169
cost effectiveness, 103, 109
COVID-19, 165
cryptocurrency, 20, 24, 26, 27, 29, 30, 34, 36, 38
cryptography, 37
currency recognition, 49, 50, 52
curriculum vitae (CVs), 145
customer behavior, 80
customer pressure, 7
cyber physical system (CPS), 97

D

data, 154, 156, 159, 162, 165, 168, 170
data analytics, 117, 118, 125
data collection, 118, 120, 121
data mining, 91
data preprocessing, 117, 118, 120, 121, 123
data reduction, 122, 123, 124
data sciences, 99
datasets, 34, 36
decision tree, 83
deepfakes, 70
deep learning (DL), 20, 40, 42, 54, 58, 95, 96,
  100, 101, 155
descriptive analytics, 118, 124, 125
diagnostic analytics, 118, 124, 125, 132
disadvantages of AI, 64, 68
disruption, 153, 164
disease detection, 42, 59

E

economic development, 96, 105
education, 156, 168–170
efficiency, 67, 70, 71, 72
electronic human resource management
  (e-HRM), 139
emerging economies, 96, 102, 110
employee involvement and empowerment, 5, 9
entertainment, 154, 159, 166, 169
environmental management (EM), 4
environmental orientation, 7
expertise, 65, 67, 71

F

facial recognition, 42, 44, 49, 50, 51, 54, 56, 57,
  58, 59
finance and accounting, 106

173

174 Index

financial decisions, 103
financial services, 167–169
forecasting, 82
Fourth Industrial Revolution, 95, 97, 98, 99

G

General Data Protection Regulation
(GDPR), 147
Generation Y, 6
green employee empowerment, 7
green human and relational capital, 7
green human resource management (GHRM), 1,
2, 3, 4, 8

H

Haar cascade, 54, 58, 59
headset, 46, 47, 48, 49
healthcare, 39, 41, 42, 43, 51, 52
HRM functions, 5
HRM practices, 5, 6
human resource (HR), 137, 138, 139, 140, 142,
143, 145, 146, 148
human resource management (HRM), 138, 139,
140, 142, 143, 148
human resource information system
(HRIS), 146
hybrid artificial intelligence (HAI), 21, 22, 23
hybrid model, 86

I

identification, 44, 51, 52, 54
individual-level approach, 85
industrial revolution, 97, 112, 154
Industry 4.0, 97, 98, 99, 100
information, 65, 66, 68, 69, 70, 71
information technology (IT), 148
infrastructure, 71
innovations, 105, 109
intelligent vision, 44, 49, 50, 51, 52
Internet of Things (IoT), 20, 43, 64, 67, 97, 99,
100, 117, 118, 119, 120, 121, 122, 124, 126,
127, 131, 132, 138, 154, 162
investigation, 87
image processing, 42, 49, 52, 53

J

job description, 5, 9

K

K-means clustering, 84
K-nearest neighbor, 87

L

labor relations, 6
lathe, 116, 117, 126
leadership, 6
ledger system, 24, 34
logistic regression, 83

M

machine intelligence, 98
machine learning (ML), 20, 37, 40, 42, 49, 50, 51,
53, 54, 101, 102, 103, 105, 106, 107, 154, 155,
159, 160, 162
machine learning technologies (MLT), 2
management by objective (MBO), 3
management by exception (MBE), 3
manufacturing, 63, 64, 65, 66, 159, 163, 164, 169
manufacturing sector, 104, 107
marketing, 71
marketing and sales, 107
media, 161, 165, 166, 168–170
metal cutting, 116, 117, 119, 120, 122, 126, 127,
129, 131, 132
microcontroller, 42, 44, 46, 47, 48, 49, 50
mining, 25, 27, 28, 159, 160, 169
mining techniques, 86
ML algorithms, 90
Mobile Aided Note Identifier (MANI), 52, 53
mobile application, 42, 48, 49
monitoring, 119, 120, 121, 122, 126, 127, 130

N

natural language processing, 3
neural network, 40, 51, 52
nodes, 23, 25, 27, 29, 33, 36
non-value-added, 119, 127, 128, 132

O

object recognition, 42, 49, 50, 51, 52
Open CV, 44, 53, 54, 58
optical character recognition (OCR), 145
optimization, 72, 73
organizational culture, 7
organizational behaviour, 106
organizational learning, 5, 9
outlier detection, 82

P

pattern recognition, 59
perception technologies, 2
performance appraisal/management, 5, 9
perspective analytics, 124, 125, 126, 132

Index

175

planning, organizing, directing, controlling (PODC), 140
precision agriculture, 160, 161, 170
predictive analytics, 118, 124, 125, 128, 129
predictive maintenance, 66, 67, 72, 121, 128, 129, 132
preventive maintenance, 128
predictive modeling, 83
privacy, 69
production, 66, 67, 71, 72, 73
production planning, 130, 131
productivity, 64, 65, 70, 71, 72, 118, 119, 120, 127, 131
Python, 51, 53, 54, 56

Q

quality, 64, 66, 68, 70, 72, 73
quality control, 131, 132

R

random forest, 83
recruitment and selection, 5, 9
reliability, 70, 71
responsiveness, 70, 71
retailing, 164
reward and pay system, 5, 9
risks, 64, 66, 67, 68, 69, 70, 72
robotics, 63, 67, 72
routinization, 71
raw brainwave data, 47, 48

S

safety, 67, 73
security, 64, 68

segment-based approach, 85
self-driving, 21, 22
sensor, 118, 119, 121, 122, 124, 127, 129, 130
setup, 127, 128, 130
skill development, 110
smart contract, 24, 30, 32, 33, 34, 37
smart manufacturing, 64, 66, 118, 119
smart technologies, 107
speech recognition, 53, 54
stakeholder pressure, 7
stoppages, 127, 128, 129, 130
supply chain, 66, 67, 72, 73
supportive climate/culture, 5, 9
support vector machine, 83
sustainable development, 155, 156
sustainability gap, 157

T

technological competencies, 97
top management support, 7
training & development, 5, 9
transaction, 23, 24, 25, 26, 27, 28, 29, 30, 32, 33, 34, 36, 37, 38
transformation, 7, 70, 153

U

unemployment, 64, 69
Union's role in environmental management, 5, 9
unsupervised algorithm, 86

V

value-added, 119, 127, 132
visually disabled patients, 42, 44, 59

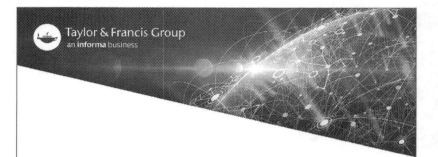

Printed in the United States
by Baker & Taylor Publisher Services